Changing Perspectives of Research with Higher Education 4.0

Challenges and Opportunities

Changing Perspectives of Research with Higher Education 4.0

Challenges and Opportunities

Editors

Dr. Upinder Dhar
Dr. Santosh Dhar
Dr. Shweta Agrawal

Shri Vaishnav Vidyapeeth Vishwavidyalaya
Indore, Madhya Pradesh, India

ALLIED PUBLISHERS PVT. LTD.

New Delhi • Mumbai • Kolkata • Chennai • Bangalore • Hyderabad

ALLIED PUBLISHERS PRIVATE LIMITED

1/13-14 Asaf Ali Road, **New Delhi**–110002
Ph.: 011-23239001 • E-mail: delhi.books@alliedpublishers.com

17 Chittaranjan Avenue, **Kolkata**–700072
Ph.: 033-22129618 • E-mail: cal.books@alliedpublishers.com

15 J.N. Heredia Marg, Ballard Estate, **Mumbai**–400001
Ph.: 022-42126969 • E-mail: mumbai.books@alliedpublishers.com

No. 25/10, Commander-in-Chief Road, Ethiraj Lane (Next to Post Office)
Egmore, **Chennai**–600008
Ph.: 044-28223938 • E-mail: chennai.books@alliedpublishers.com

P.B. No. 9932, No. 15, 3rd Floor (Next to Vijaya Bank), 5th Cross, Gandhinagar, Karnataka, **Bangalore**–560009, Ph.: 080-41530285 / 22386239
• E-mail: bngl.journals@alliedpublishers.com / apsabng@airtelmail.in

Sri Jayalakshmi Nilayam, No. 3-4-510, 3rd Floor (Above More Super Market)
Barkatpura, **Hyderabad**–500027
Ph.: 040-27551811, 040-27551812 • E-mail: hyd.books@alliedpublishers.com

Website: www.alliedpublishers.com

ISBN: 978-93-87997-53-0

Prologue

Paradigm Shift in Knowledge Creation

Entrepreneurship is a way of thinking that emphasizes on opportunities instead of threats. The identification of opportunities is a process based on intentions. Entrepreneurial activity is a direct result of an "individual's perception about the existence of market opportunities, capacity and economic viability of those business opportunities." Entrepreneurship corresponds to a situation where an individual believes that an opportunity could provide higher returns than an occupation, or when in an unemployment situation the need becomes a vital drive to solve the problem.

Higher education institutions are required to demonstrate the ways in which they respond to the social and economic needs of society, such as enhancing graduate employability, facilitating social mobility and wider access to higher education, contributing to national economic growth and local development, stimulating new enterprises and innovation in existing firms. In addition, higher education institutions must continuously adapt and respond to new challenges to maintain standards of excellence and be competitive in international education markets. Higher education institutions behave in an entrepreneurial and innovative manner in a number of ways.

For example, how they manage resources and build organisational capacity; involve external stakeholders in governance; embed digital technology into their activities; create and nurture synergies between teaching, research and their societal engagement, and how they promote entrepreneurship through education and business start-up ecosystem as well as knowledge exchange to improve the capacity of existing firms for innovation.

Entrepreneurship as a process can promote change and development by enhancing the capacity to recognise and act upon opportunities. As such, entrepreneurship has a long-standing presence in higher education reform initiatives, promoting, for instance, the interdisciplinary approach in teaching and research, and in engaging external stakeholders for enriching the academic offerings with pragmatic and proactive touch.

It depends on individuals and innovative ways of doing things, and a supportive organizational culture to nurture entrepreneurial ecosystem in a higher education institution. Promoting entrepreneurial higher education is not about re-labelling, it is about recognizing and building – in innovative ways – on what already exists.

As one of the consequences of the globalisation and digital transformation, borderless education has been a key enabler for the paradigm shift in knowledge creation. No single university can any longer claim to be the principal repository of knowledge. Independence of academic discovery and teaching processes is still widely present along with the assumption that higher education institutions are:

- sustained by a mode of thought which is shared by all of their stakeholders,
- yet underpinned by a detachment of their members' motivations from the goals and functions of the organization.

Such independence has in many countries traditionally been supported by government funding based on the quality and extent of research and publications. Excellence has been viewed through the lens of peer review processes, and particularly through the prism of publications in high-impact journals, which is now being challenged. Higher education institutions, in general, are increasingly directed to enhance their capacity to focus on:

- useful problem-centred sources of knowledge, and create wider partnerships for learning,
- cross disciplinary boundaries, and discover, exploit, and share knowledge in new ways.

The digital transformation will bring significant changes to teaching and learning in higher education. Teachers and students shall have to cope with the sheer volume of information which is freely available on the Internet, condensed and presented appealingly to students, enabling them to easily go beyond the recommended readings.

Academic blogs, YouTube, Facebook, Twitter, and other forms of social media have gained growing significance in teacher-student communication. Acting as new channels of learning, they question traditional approaches to teaching. One reaction to this is flipped classrooms: students are asked to 'discover' more of their learning and use conventional lectures from online sources.

Students are challenged to become aware of a wider range of knowledge sources and find novel solutions, whereas the teacher becomes a facilitator of learning. The potential for wider student learning has further been strengthened by the growth of Massive Open Online Courses (MOOCs), attracting large number of students worldwide. Developments in and use of digital technologies provide opportunities for innovative curriculum design and delivery, and it also enables new ways of pursuing and assessing progress.

Responding to these challenges and opportunities encompasses a different approach to teaching. It requires rethinking of the education mode and a significant organizational innovation effort. Integrating various sources of information and knowledge together into a dynamic and open learning environment leads to the situation where teachers and students interact, reflect and create knowledge, while ensuring interdisciplinary and flexible study programs.

These developments are stimulating the spread of virtual academic institutions, bracing, in turn, the need for short-term study arrangements, flexible modes of credit accumulation and mobility between institutions. Internationalization offers valuable benefits to students, faculty, and the institution as a whole. International mobility of faculty and students enhances academic enrichment through exposure to new research environments and application opportunities.

A widely practiced approach to internationalization is setting up partnerships with higher education institutions abroad to facilitate virtual and physical faculty and student exchanges, collaboration in research and development, international joint degree programs and the opening of campuses abroad. Opening up wider links through distance learning approaches, globalization of curricula, building stronger linkages with local international businesses and closer engagement with alumni abroad are also viable practices.

This book has put together some of the papers received for *SHODH-2021* after thorough editorial inputs and necessary revision whenever required. We are confident that delegates and readers will find contents of this book useful for understanding the orientation of young research scholars of the contemporary times. One book cannot be representative of the whole mass of research that is pursued at present, but it certainly provides a glimpse of the research environment in the country.

Date: 09.03.2021

Dr. Upinder Dhar
Dr. Santosh Dhar

About the Editors

Dr. Upinder Dhar is Vice Chancellor of Shri Vaishnav Vidyapeeth Vishwavidyalaya (Indore). Earlier, he was Group Additional Vice Chancellor and Pro-Vice Chancellor (Academics), Amity University (Noida); Founder Vice Chancellor of JK Lakshmipat University (Jaipur); Director - Institute of Management, Dean - Faculty of Management, and Dean - Faculty of Doctoral Studies and Research, Nirma University (Ahmedabad); Director - Prestige Institute of Management and Research (Indore); Professor of Management at NITIE (Mumbai); Reader at IMS-DAVV (Indore); and Lecturer at M.D. University (Rohtak). He has conducted more than 650 management and faculty development programs for the executives of major business houses and faculty members of academic institutions respectively. He has also been Guest Faculty at IIM Calcutta, IIM Indore, IIM Trichy and IIT Roorkee. Besides undertaking consultancy assignments, he has authored/co-authored/edited more than 700 publications. Thirty-Six (36) scholars have been awarded Ph.D. degree under his supervision by various universities.

A Fellow of ISTD and a strong proponent of Case Method, Dr. Dhar was conferred **Ravi J Matthai National Fellow Award** by the *Association of Indian Management Schools (AIMS) in 2018.* He has been on the Panel of Reviewers of number of international journals, such as International Journal of Training and Development, Human Resources International, Journal of Global Business and Technology, and South Asian Journal of Management. He has also been on the Panel of Experts maintained by the NAAC (Bangalore), NBA (Bangalore); ICSSR (New Delhi); and SAQS (Hyderabad). He was President of Anthropological and Behavioural Sciences Section of the Indian Science Congress Association in 2007–08, Indian Society for Training and Development in 2008–09 and Association of Indian Management Schools in 2014–15. He has also been Member of the Council of Management of AIMA, Member of ATMA, All India Board of Management Studies of AICTE, Rajasthan State Productivity Council, Governing Board of DELNET, and Member of the Executive Board of AMDISA.

Dr. Dhar was awarded Commonwealth Fellowship twice to present his papers at the South Asian Management Forums held at Royal Institute of Management, Bhutan and Lahore University of Management Science, Lahore. He was invited by the American Society for Engineering Management as resource person for Case Method during its international annual Conference held at Minnesota in 2013. He had the opportunity to visit Ivy league institutions like Harvard Business School, MIT Sloan School of Management, and Yale University School of Management. He has also visited 18 countries for professional and academic interest. He has been on the high powered Expert Committees constituted by MHRD, UGC, AICTE and Hon'ble Governor of Uttar Pradesh. Currently, he is a Member of the *Standing Committee on Specification of Degrees u/s 22* of the *UGC Act, 1956*. He has also been resource person in the Leadership for Academicians Program (LEAP) sponsored by MHRD and organized by IIT Roorkee in 2018 and BHU Varanasi in 2019.

Dr. Santosh Dhar is Dean-Faculty of Doctoral Studies and Research at Shri Vaishnav Vidyapeeth Vishwavidyalaya, Indore. She has been engaged in teaching, research, corporate training and consultancy for the last forty years. She has widely published her work and has more than 356 publications to her credit. Her earlier assignments have been Professor, Shri Vaishnav Institute of Management, Indore; Professor, Amity Business School, Amity University, Greater Noida campus; Professor, Institute of Management and Dean - Research, JK Lakshmipat University, Jaipur; Senior Professor, OB/HR, Jaipuria Institute of Management, Jaipur; Professor and Chairperson - OB/HR Area, Institute of Management, Nirma University, Ahmedabad and Guest Faculty at IIM Kolkata and MICA Ahmedabad. Many research scholars have been awarded Ph.D degree under her supervision. She has designed and conducted management development programs for the executives of several business houses, such as Grasim Industries, JK Tyre, Bridgestone, Hindustan Motors, ONGC and GAIL to name a few. She has been on the Panel of Reviewers of Oxford University Press and Tata McGraw-Hill. She was Managing Editor of Prestige Journal of Management and Research and Prestige Research Abstracts. She has been Chairperson of Indian Society for Training and Development (Indore and Jaipur Chapters) and was on the Editorial Board of Indian Journal of Training and Development. She has presented papers in several national and international conferences/seminars and had the opportunity to visit Kathmandu (Nepal), Male (Maldives), Bangkok (Thailand), Beijing, Shangai, Guanzhou, Shenzen, Hong Kong (China), Seoul (South Korea) Dubai (UAE), Tumu (Myanmar), Poland and United States in connection with various professional and academic assignments.

Dr. Dhar had coordinated the implementation of a project on the *Empowerment of Women through Enterprise* in North India (2000–2002), which was funded by Canadian International Development Agency (CIDA). She has completed the research projects on *Problems of Women Trainers in India* and *Quality of Work Life as a Determinant of Mental Health in Pharmaceutical Industry*. She has also been principal investigator for Asian Development Bank funded and Rajasthan Government sponsored RUIDP Consultancy project on *Capacity Building of ULBs* of Different Towns of Rajasthan. She has twice been chairperson of Indian Society for Training and Development Jaipur and Indore Chapter & Chairperson of the Best Paper Award of the Indian Society for Training and Development, New Delhi. Dr. Dhar was conferred *National Fellowship* for her contributions to HR by the Indian Society for Training and Development in 2006–07. She is life member of Indian Society for Training and Development (ISTD), National Human Resource Development Network (NHRDN), and Indian Society for Technical Education (ISTE).

Dr. Shweta Agrawal is Assistant Professor at the Department of Life Science, Shri Vaishnav Institute of Science, Shri Vaishnav Vidyapeeth Vishwavidyalaya, Indore, M.P, India. She has been engaged in teaching and research for nearly 14 years with specialization in the area of Microbiology and Environmental Biotechnology and IPR. She has taught IPR at national and International universities having Indian, Nepali and Bhutanese students.

She is life member of Association of Microbiologists of India, Biotech Research Society of India and Microbiologist's society of India. She is a recipient of Best Thesis award by GSA-CHARUSAT (2014), Best paper presentation award at the International conference organized by Janparishad and MPCoST Bhopal (2014) and Best Teacher Award by Microbiologist's Society, India.

She has more than 20 publications in journals of national and international repute. She has participated in a number of workshops, faculty development programmes, national and international conferences. Dr. Agrawal has been the resource person for National Level Workshops on Gene cloning, Career counselling organized/sponsored by MPCoST, BRSI, MBio Society. She is also a member of the editorial board and reviewer for many journals.

Contents

A Study on Security and Privacy of Smart Contract using Blockchain Technology

Jitendra Sharma[1] and Jigyasu Dubey[2]

ABSTRACT

Blockchain is novel and effective technology, which facilitates the implementation capability of various areas like public services, economic systems, Internet of Things, Internet interaction systems and Smart Contract applications. Smart Contract works in such a way that the computer works as a third party, and the other two senders and receiver are those who digitally agree to the obligations and rights. Blockchain keeps the record of the Contract which are executed by the network's distributed nodes, where the third-party requirement is eliminated as Blockchain decentralized in nature. There are still multiple security and privacy issues with the Smart Contract which impose a huge threat for its continued effectiveness and growing esteem and recognition. The present paper, addresses the issues of transactions in smart contracts.

Keywords: Blockchain, Privacy, Security, Smart Contract.

INTRODUCTION

In the search for more effective and reliable software products, the modern advancements in the technological arena are frequent in bringing in novel technologies. The introduction of one such novel and effective technological discovery is Blockchain, which renders the implementation capability of various cases like public services, economic systems, Internet of Things and Internet interaction systems. The industry and academia have recognized the importance of Blockchain Technology (Zheng *et al.*, 2017). In Blockchain Technology, the cryptography is utilized to make a transfer of digitalized currency/messages, in hosting applications in a secured and protected manner and store data (Dannen *et al.*, 2017). This makes Blockchain a completely-distributed-public-ledger and a peer-to-peer platform. The Blockchain invention is considered remarkable in the world of INTERNET because it can make faster processes, lessen the cost of

1 Research Scholar, Shri Vaishnav Institute of Information Technology, Shri Vaishnav Vidyapeeth Vishwavidyalaya, Indore, Madhya Pradesh, India. E-mail: js1102@outlook.com

2 Professor and Head, Shri Vaishnav Institute of Information Technology, Shri Vaishnav Vidyapeeth Vishwavidyalaya, Indore, Madhya Pradesh, India. E-mail: jigyasudube@yahoo.co.in

transactions, and ultimately bring effectiveness in many sectors (Drescher *et al.*, 2017).

The Blockchain technology is mainly used for cryptocurrencies; however, it may be useful in applications in which virtual records maintainance is required like government issued certificates, medical procedures, deeds of ownership, educational certificates, election, Smart Contract etc. Researches are attracted to use Blockchain in Smart Contract applications because it works in such a way that the computer works as a third party and the other two are those who digitally make an agreement upon the obligations and rights. Smart contract are programmed in unique codes. The variables are ordered as codes to trigger various events. On agreeing the terms and conditions the contract is signed digitally between parties.

After this, the Blockchain keeps the record of the contract and execute it by the network's distributed nodes, where the third party requirement is eliminated. These contracts are imposed on parties and they are carried out just as they were coded. The Blockchain ledger is transparent and facilitates variables that are required for the contract. It triggers events in smart contract like making a payment and unlocking the doors for the Airbnb house. Smart contracts have different issues like codifying, security, privacy, and performance. The codifying issue is related to the development of smart contracts. Security issues are related to bugs or vulnerabilities that an adversary might utilize to attack. The privacy issue is related to disclosing contact information in public. The performance issue is related to the scalability of the Blockchain system.

EFFECT OF BLOCKCHAIN AND SMART CONTRACTS

Blockchain Technology is not only a string of static data records stored in blocks. But also ca execute computer programs to perform functions. The Smart Contract (SC) is a code embedded into the Blockchain and contains all or part of the legal agreement and transfers the digital assets. It is also defined as SC which is a self-triggered code and will be executed when a set of predefined terms and conditions are satisfied. The SC once written can be self-executing and autonomous, without the need of further actions by parties. The Blockchain can ameliorate all the features of the government. The huge bureaucracy and massive scale make room for improvement in government organizations. It is desired to make the services quicker, reliable, and efficient.

There is a tremendous chance of replacing certain elements of the government by the Blockchain. The developments by Blockchain can affect the government positively by ameliorating the abilities of taxes, control over interactions in the economic sector, the privacy policy, and even the status of the government. To find suitable smart contracts for implementation in the government services, it is required to look at the present cases of implementation beforehand. The use of smart contracts in government services is discussed in this research (Feng *et al.*, 2019). The implementation is categorized into six divisions: Supply-chain, Identification, Voting, Ownership, Financial, and Permit Application.

REVIEW OF LITERATURE

The paper contracts have become outdated. For a long time, contracts have been concluded and stored digitally. With a due delay, the law has accepted the notion that digital contracts are valid even where certain formal requirements must be met. In the present scenario, digital contracts are part of the legal mainstream. Smart contracts are designed as a mechanism that acts as an efficient and trusted middleware in the financial transactions. The term smart contract was coined by Nick Szabo in 1994. He defined a smart contract as a computerized transaction protocol that executes the terms and conditions of a contract. The explanation of the fundamental of the technology of Blockchain is discussed: a ledger of a distributed type that keeps the transaction records. This is Public-Blockchain because they are available for everyone to view and verify. For interacting within the network, there is no need to take any permission. So, it is permission-less (Tapscott *et al.*, 2016). There is one more form of this sort called Private-Blockchain. In this, the transaction records are arranged and only open to permission.

The Blockchain is a distributed database that records transactions that occurred in the Blockchain network. Cryptography is the root of the Blockchain technology. The Blockchain concept is behind the Bitcoin. It describes that every single party should have two different keys: a private key and a public key. The private key is the password, whereas the public key is the account number. The basic overview of the Bitcoin is shown in Figure 1. The private key is used whenever the transaction has to be made and sent to the server of Time-stamp for endorsement. With many transactions, a kind of block is filled. Cryptographically, these blocks are hashed, just like the other previous blocks. The history of all the transactions are correct, is proved by this hash. The chances of forfeiting history are very thin because of the hash functions of cryptography (Nakamoto, 2008, Kosba *et al.*, 2016). Summarizing, in Bitcoin technology, the transaction can only be completed via private key, the dual-spending is restricted, and the records of all the history is available.

The counterparty is not required to be trusted while making any transactions because the verification is done by the network. Therefore, the word "Trustless" is used for the Blockchain. The third-party, like, Bank is not required in these processes. According to Swan *et al.* (2015) Work Proof of Stake (Vukolić *et al.*, 2015) and BFT Replication (Iansiti *et al.*, 2017) also exists. There is no discussion of these methods in this research as they are dealt in the technical level.

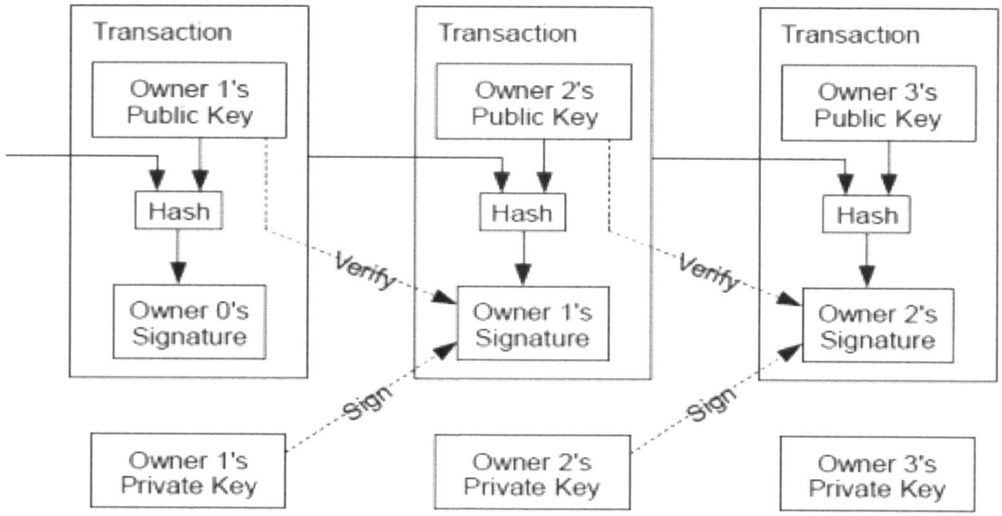

Fig. 1: Hashing Scheme of Bitcoin [7]

The transaction history is recorded and verified by the decentralized network. All the nodes in the network are equally important because transmission of information occurs in all the nodes that are participating. Vanini *et al.* (2018) called this as peer-to-peer transmission. To be accurate, the hash requires the whole history, hence, the complete database must be kept as possession by the hash whenever novel blocks are mined. The hacking of the database becomes impossible as all the nodes that are participating have the whole copy.

As an attribute, the transparency is brought by the distributed datasets. Because of the Public Blockchain, which explains that each node possesses the block chain's copy, all the transactions are visible to everyone, and recent histories are shared as to whom and where the transactions did take place (Swan *et al.*, 2015). The pseudonyms of the address protect the transparency. Each user or node is assigned with an address of the Blockchain. These addresses contain 30 or above alphanumeric characters. This provides privacy of the address, but in case if it is required, the identity can be proved by transacting (Vanini *et al.*, 2018). In conclusion, we can say that transparency is provided to the transactions, but the pseudonyms keep the users private and protected.

The Blockchain data is immutable because of the distributed database and the basic fundamentals of cryptography. By the utilization of the preceding blocks, hashing of novel blocks is done. All the used blocks must be corrected to produce novel blocks. If someone indulges in the Bitcoin theft, he has to rephrase the whole history of the transaction, which is actually unattainable. It is not what happens if someone attempts to alter the complete record. The pioneering Bitcoin of the Blockchain technology is unhackable (Barcelo *et al.*, 2017).

Therefore, the Blockchain records are thought to be totally immutable. There are some securities issues present on the Blockchain. 51% of attacks are considered very dangerous. When nodes in the network vote 51 or more percentage maliciously, there are chances that the transactions are falsely marked as correct, and the history is altered (Swan *et al.*, 2015). One more major issue of security is that the applications sometimes have bugs because of the newly applied characters. Rather than just these bugs, the hackers eye the individual's crypto-currencies.

The transparency of ledger is the main feature of Blockchain. As many of the details of transactions are open to all, the issues of privacy are always implied, e.g., the transfer details of cryptocurrency are visible as time, and the address of both sender and recipient is made public (Drescher *et al.*, 2017). There are also a hand-full of tools of Flow-Control that are available to show the transactions which are done in the past. Natoli *et al.*, (2016) researched that the identities of various users' accounts are obtainable. The Wiki-leaks organization has first done this. The addresses of owners were mentioned on the website of Wiki-leaks with the objective of receiving the donations. Because the transfer of Bitcoin amounts is shown by the Blockchain. So, it became obvious that some people could merge their identities with some contributors' addresses.

Blockchain technology has developed since 2008. The applications that work on the Blockchains are hacked, but the Blockchain as a technology in itself is free of such accusations. The guidance of Industry standards is required to be guided in designing safer and adoptable technologies (Zhang *et al.*, 2016).

The standardization in the Blockchain technology was requested by the ISO (International Organization of Standardization) in Australia (Silverberg *et al.*, 2018). Along with 25 members, the ISO is working in the development of the Blockchain standards. The members are from various other countries like France, Brazil, Germany, Netherlands, United Kingdom, Russia, the United States, Japan, China, and Australia. The code for this standard is ISO/TC 307, and it is in the preliminary stage. There are so many smart contracts that are completely reliant upon the values of external affairs. e.g., the Contract of betting depends upon the results of a football match. The inputs of the smart contracts are given by a source called Oracle (Delmolino *et al.*, 2016). In the example of football betting, n Oracle can be considered as the organization that released the outcomes of a match. In oracle, the third-party is involved as an intermediate, which is against the rules of Blockchain (Hooghe *et al.*, 2018). The illustration of a smart-contract by means of an example of "Financial-swap" is given in Figure 2.

Let us take the example of Bob and Alice who agree to a Financial-swap. The intention is to make a bet over the stock prices. Bob says that the stock-price of the micro-soft will end on $85 or can be lower than that, au contraire; Alice wishes to

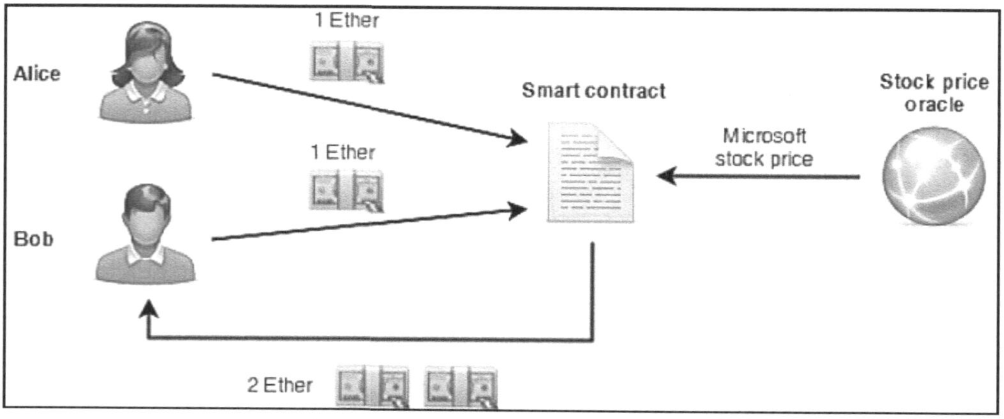

Fig. 2: Smart Contract Example

bet on the increment of that amount. As per the rules of smart-contract, they both have to make the deposit of 'one' currency of Ether. With the help of their private and public keys, they have to sign the deal. The smart-contract is executed after a given deadline. The Stock Price Authority queries the stock price details. This authoritative organization is an oracle (Atzori *et al.*, 2018). Feng *et al.* (2019) provided information about the different privacy issues related to Blockchain system. It gives brief introduction about different types of Blockchain and technologies which can resolve this issue. It also discussed about different types of currencies. NIZK (Non-Interactive Zero Knowledge), Homomorphic Cryptosystem for preserving Privacy of Blockchain System.

Singh *et al.* (2019) showed that the most common formalization technique is theorem proving, which is most often used to verify security properties relating to smart contracts, while other techniques such as symbolic execution and model checking were also frequently used. These techniques were most used to verify the functional correctness of smart contracts. It highlights three open issues for future research in this emerging domain including: formal testing, automated verification of smart contracts, and (DSLs) for Ethereum. Unterweger *et al.* (2018) provided implementation of privacy preserving smart contract using Ethereum and showed that it is not feasible to implement a privacy-preserving protocol of modest complexity in the Ethereum Blockchain due to high operational cost and lack of privacy in design. It is required more GAS and unable to handle large amount of data. It delays in transaction to confirm.

Blockchains and smart contracts lack standardization. Therefore, smart contracts come with heterogeneous properties, APIs and data formats. This hinders the integration of smart contracts running in different blockchains, e.g., into enterprise business processes. It introduces the Smart Contract Invocation Protocol (SCIP), which unifies interacting with smart contracts of different Blockchains (Falazi *et al.*, 2020). The protocol supports invoking smart contract functions, monitoring function executions, emitted events, and transaction finality, as well as querying a blockchain.

CONCLUSION

The smart Contract and Blockchain plays a vital role in changing the current technology and trends. The paper highlighted the issues of smart Contracts like codifying, security, privacy, and performance and related tools that are currently in use to overcome these issues. Our focus in this literature survey on the privacy and security of the transaction performed smart contract over Blockchain.

REFERENCES

[1] Alharby, M. and van Moorsel, A. (2017). Blockchain-based smart contracts: A systematic mapping study. *arXiv preprint arXiv:1710.06372.*

[2] Atzori, M. (2018). Blockchain Technology and Decentralized Governance: Is the State Still Necessary? Retrieved online from https://papers.ssrn.com/sol3/papers.cfm?abstract_id=2709713.

[3] Barcelo, J. (2017). User Privacy in the Public Bitcoin Blockchain. *Journal of Latex Class Files*, 6(1).

[4] Chatzopoulos, D., Gujar, S., Faltings, B. and Hui, P. (2018). Privacy Preserving and Cost Optimal Mobile Crowd sensing using Smart Contracts on Blockchain. *15th International Conference on Mobile Ad-hoc and Sensor Systems.*

[5] Coinmarketcap.com (2017). Crypto Currency Market Capitalizations. Retrieved online from https://coinmarketcap.com.

[6] Dannen, C. (2017). Introducing Ethereum and Solidity: Foundations of Cryptocurrency and Blockchain Programming for Beginners. *Apress, Berkely, CA, USA,* 1st edition, 10.1007/978-1-4842-2535-6.

[7] Delmolino, K., Arnett, M., Kosba, A., Miller, A. and Shi, E. (2016). Step by Step Towards Creating a Safe Smart Contract: Lessons and Insights from a Cryptocurrency Lab. *International Conference on Financial Cryptography and Data Security,* pp. 79–94, Christ Church, Springer.

[8] Drescher, D. (2017). Blockchain Basics: A Non-Technical Introduction in 25 Steps. *Apress,* ISBN: 9781484226049.

[9] EC (2017). Blockchains for Social Good. https://ec.europa.eu/research/eic/index.cfm?pg=prizes_blockchains.

[10] Falazi, G., Breitenbü¨ucher, U., Daniel, F., Lamparelli, A., Leymann, F. and Yussupov, V. (2020), Smart Contract Invocation Protocol (SCIP): A Protocol for the Uniform Integration of Heterogeneous Blockchain Smart Contracts, *CAiSE 2020, LNCS 12127,* pp. 134–149, https://doi.org/10.1007/978-3-030-49435-3_9

[11] Feng, Q., He, D., Zeadally, S., Khan, M.K. and Kumar, N. (2018). A survey on privacy protection in blockchain system. *Journal of Network and Computer Applications,* doi: https://doi.org/10.1016/j.jnca.2018.10.020.

[12] Henry, R., Herzberg, A. and Kate, A. (2018). Blockchain Access Privacy Challenges and Directions. *Co-published by the IEEE Computer and Reliability Societies.*

[13] Iansiti, M. and Lakhani, K.R. (2017). The Truth About Blockchain: Harvard Business Review. 95(1), 118–127.

[14] Kosba, A., Miller, A., Shi, E. and Wen, Z. (2016). Hawk: The Blockchain Model of Cryptography and Privacy-Preserving Smart Contracts. *IEEE Symposium on Security and Privacy* pp. 1–20.

[15] Luu, L., Chu, D.H., Olickel, H., Saxena, P. and Hobor, A. (2016). Making Smart Contracts Smarter. *ACM.* ISBN 978-1-4503-4139-4/16/10, pp. 1–16.

[16] Nakamoto, S. (2008). Bitcoin: A peer-to-peer electronic cash system. *bitcoin.org.*

[17] Natoli, C. and Gramoli, V. (2016). The Blockchain Anomaly. *arXiv:1605.05438v1* pp. 1–11.

[18] Silverberg, K., French, C., Ferenzy, D. and Berg, V.D.S. (2018). Getting Smart Contracts on the Blockchain. *Institute of International Finance. Retrieved online on the 4th of March 2018 from* https://www.iif.com/system/files/32370132_smartcontracts_report_may_2016_vf.pdf.

[19] Singh, A., Parizi, R.M., Zhang, Q., Choo K.K.R. and Dehghantanha, A. (2019). Blockchain Smart Contracts Formalization: Approaches and Challenges to Address Vulnerabilities. *Computers & Security Elsevier*, doi: https://doi.org/10.1016/j.cose.2019.101654.

[21] Swan, M. (2015). Blockchain: Blueprint for a New Economy. *O'Reilly Media.*

[22] Unterweger, A., Knirsch F., Leixnering, C. and Engel, D. (2018). Lessons Learned from Implementing a Privacy-Preserving Smart Contract in Ethereum. *NTMS'18*,978-1-5386-3662-6/18/$31.00 ©2018 IEEE.

[23] Vanini, P. (2018). Digital Disruption. Available at SSRN: https://ssrn.com/abstract=2988962 or http://dx.doi.org/10.2139/ssrn.2988962 Vukolić, M. (2015). The Quest for Scalable Blockchain Fabric: Proof-Of-Work vs. BFT Replication. *International Workshop on Open Problems in Network Security* pp. 112–125, Springer.

[24] Zhang, F. (2016). Town Crier: An Authenticated Data Feed for Smart Contracts. *ACM* pp. 1–20.

[25] Zheng, Z., Xie, S., Dai, H.-N. and Wang, H. (2017). Blockchain challenges and opportunities: A survey. *International Journal Web and Grid Services.*

Advanced Conducting Material

Polyaniline (PANi)-Copper Oxide (CuO) Nanocomposites as Sustainable Energy Resources

Supriya Vyas[1], Navneeta Upadhyay[2], Shweta Agrawal[3] and Ashutosh Shukla[4]

ABSTRACT

The aim of present work is to fabricate the composite of polyaniline with copper oxide nanoparticles which is applicable in many sustainable energy resources due to its high conducting nature. The Cupric oxide (CuO) nanoparticles were synthesized by chemical co-precipitation method and the pure Polyaniline (PANi) and PANi-CuO nanocomposites were chemically synthesized using in-situ oxidative process. The structural determination was carried out by using the techniques such as X-ray Diffraction (XRD) studies, Fourier Transform Infrared Spectroscopy (FTIR) and UV-Visible Absorption Spectrometry (UV-Vis Spectroscopy). The peaks obtained in spectra validate the fabrication of desired materials. The morphology of synthesized materials was explored by Scanning Electron Microscopy (SEM). The conductivity of materials was examined by using Four–Probe Technique, which confirms the semiconducting nature of composite. The antibacterial activity of composites was performed by disk diffusion method using Escherichia coli, (Gram negative bacteria) and Bacillus subtilis (Gram positive bacteria).

Keywords: Conducting Material, Co-precipitation Method, Four-probe Technique, Disk Diffusion Method.

INTRODUCTION

Intrinsically conducting polymers (ICPs) is an emerging field in material science due to the interesting electrical and optical properties of these polymers. PANi is presently a very recognizable polymer among the conducting

[1]Research Scholar, Shri Vaishnav Institute of Science, Shri Vaishnav Vidyapeeth Vishwavidyalaya, Indore, Madhya Pradesh, India. E-mail: supriyavyas2012@gmail.com

[2,3]Assistant Professor, Shri Vaishnav Institute of Science, Shri Vaishnav Vidyapeeth Vishwavidyalaya, Indore, Madhya Pradesh, India. E-mail: [2]navneetaupadhyay@gmail.com; [3]shweta.agrawal24@gmail.com

[4]Professor, Shri Vaishnav Institute of Science, Shri Vaishnav Vidyapeeth Vishwavidyalaya, Indore, Madhya Pradesh, India. E-mail: drashutoshshukla@yahoo.co.in

polymers because of its unique properties and its applications, good environmental stability, easy synthesis and low cost (Ashokkumar *et al.*, 2017). A chemically flexible-NH group has attached on both the sides of polymer PANi with phenylene ring (Vyas *et al.*, 2017). Over some drawbacks such as chemical swelling, shrink, solubility and weak mechanical properties, the various applications of PANi may reduce. Metal oxides dispersed polymer composites reveal unique blended properties derived from both the constituents (Nadaf *et al.*, 2015). The semiconducting nature of CuO makes CuO-PANi nanocomposites applicable in sustainable energy resources like batteries, sensors (Shu'aibu *et al.*, 2018), solar cells, supercapacitors (Ashokkumar *et al.*, 2020) and antimicrobial (Singh *et al.*, 2020) applications.

The development of materials which can minimize or prevent infectious microbial colonization is an urgent requirement to reduce diseases that affect public health and global economies (Tamayo, 2016). The CuO nanoparticles embedded in PANi matrix also prove to have good antimicrobial activity (Elango *et al.*, 2017). In the present study CuO nanoparticles, PANi and PANi-CuO nanocomposites were synthesized with *in-situ* chemical oxidative polymerization. The structural and morphological analysis of fabricated materials were carried out by XRD, FTIR, UV-Vis spectroscopy and SEM. The measurement of conductivity was performed by Four-probe Technique. Additionally, the Disk Diffusion method was used to determine the antibacterial properties.

MATERIALS AND METHOD

Synthesis of Cupric Oxide (CuO) nanoparticles

The fabrication of cupric oxide nanoparticles were accomplished by chemical co-precipitation method. Firstly, cupric nitrate trihydrate solution (0.2M) was prepared in double distilled water. Copper hydroxide gel was formed on dropwise addition of ammonia in that cupric nitrate trihydrate solution which was continuously stirred for 8 hours at 85 ± 2°C. The black shiny CuO crystals thus obtained was washed with distilled water and ethanol then filtered and dried in oven.

Synthesis of Pure PANi

PANi was fabricated by using the method oxidative polymerization of aniline. In this method two types of solutions were prepared: (a) 0.2M aniline solution in 100 ml 1M HCl and (b) 0.2M ammonium persulphate solution in 50 ml distilled water. The oxidant based solution was added to the aniline based solution dropwise with constant stirring on magnetic stirrer at 0 ± 5°C. The solution turned green from colourless. Now, the solution was further stirred for 5 hours using magnetic stirrer at 0 ± 5°C and then placed in a refrigerator for 12 hours.

The green precipitates were first washed with distilled water and then with acetone. The synthesized green material was dried in oven at 60°C and grounded to fine powder using mortar and pestle.

Synthesis of PANi-CuO Nanocomposites

The nanocomposites of PANi and CuO nanoparticles were amalgamated by *in-situ* chemical oxidative polymerization process using HCl as dopant and ammonium persulphate as oxidant. In this process, 0.2 M aniline solution was prepared in 100 ml of 1 M HCl. In this aniline monomer solution CuO nanoparticles were added and stirred the solution in an ice-bath for half an hour. 50 ml of 0.2 M ammonium persulphate solution was added dropwise to the above solution and stirred at low temperature. Now, the solution further stirred for 5 hours using magnetic stirrer at 0 ± 5°C and then placed in a refrigerator for 12 hours. The product formed was washed with acetone and distilled water and dried in oven at 60°C.

Characterization

The X-ray Diffraction (XRD) patterns were recorded on a Bruker's model D8 advance system using nickel filtered Cu-Kα radiation as the X-rays source (λ = 1.54178 Å). The average particle size of synthesized materials was measured by using Scherrer formula. The UV-Vis spectra were recorded in the wavelength range of 200–800 nm and the FTIR spectra were recorded in the frequency range of 400–4000 cm^{-1}. The JEOL Scanning Electron Microscope was used for morphological studies of synthesized materials. The Four-probe conductivity measurement technique was used to calculate the DC conductivity of synthesized compounds. The antibacterial activity nanocomposite was determined by disk diffusion method with concentrations of 5 mg/ml and 10 mg/ml against *Escherichia coli* (Gram negative bacteria) and *Bacillus subtilis* (Gram positive bacteria).

RESULTS AND DISCUSSION

X-Ray Diffraction (XRD) Studies

Figures 1, 2 and 3 show the XRD patterns of pure polyaniline (PANi), Copper oxide (CuO) nanoparticles and PANi-CuO nanocomposites, respectively. Samples were step scanned in terms of 2θ in the range 20° to 80° angle. The pure PANi (Figure 1) shows a diffraction peaks of 2θ = 20.07° which attributes to the characteristic interval among the planes of benzene rings in adjacent chains (Zhou *et al.*, 2014). The peak obtained at 2θ = 24.92° can be allocated to the dispersion of PANi chains at the interplanar spacing (Ge *et al.*, 2012, Gulce *et al.*, 2013). The less intense peaks support the amorphous nature of PANi.

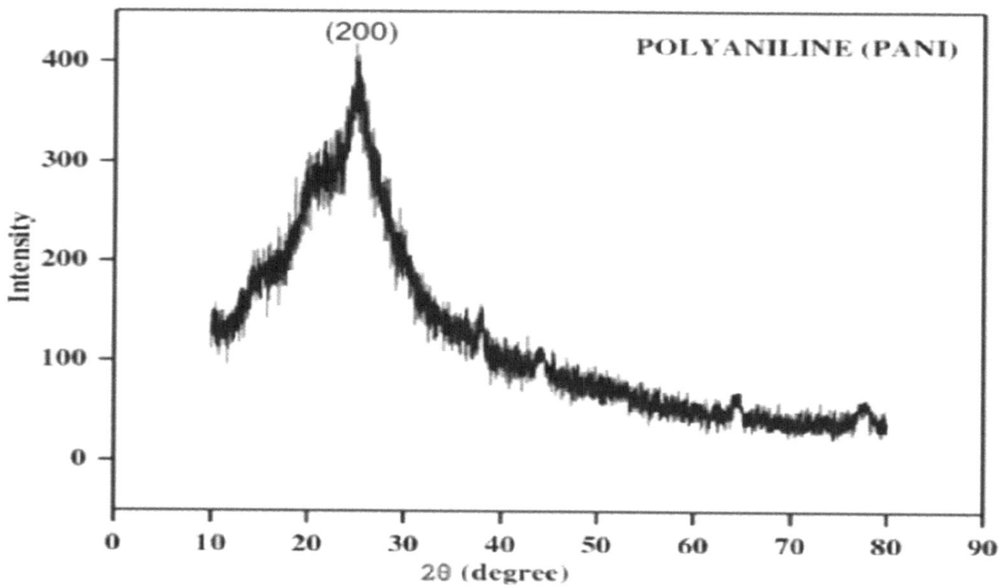

Fig. 1: X-Ray Diffraction Pattern of Pure PANi

Figure 2 exhibits the XRD pattern of nanoparticles of CuO in which the sharp peaks indicating the high crystalline nature of nanoparticles. Sharp peaks are observed at 2θ = 35.6° and 38.7° which are in good agreement with the reported values (JCPDS card number 89–5899). The broad peaks indicate the nano-size effect.

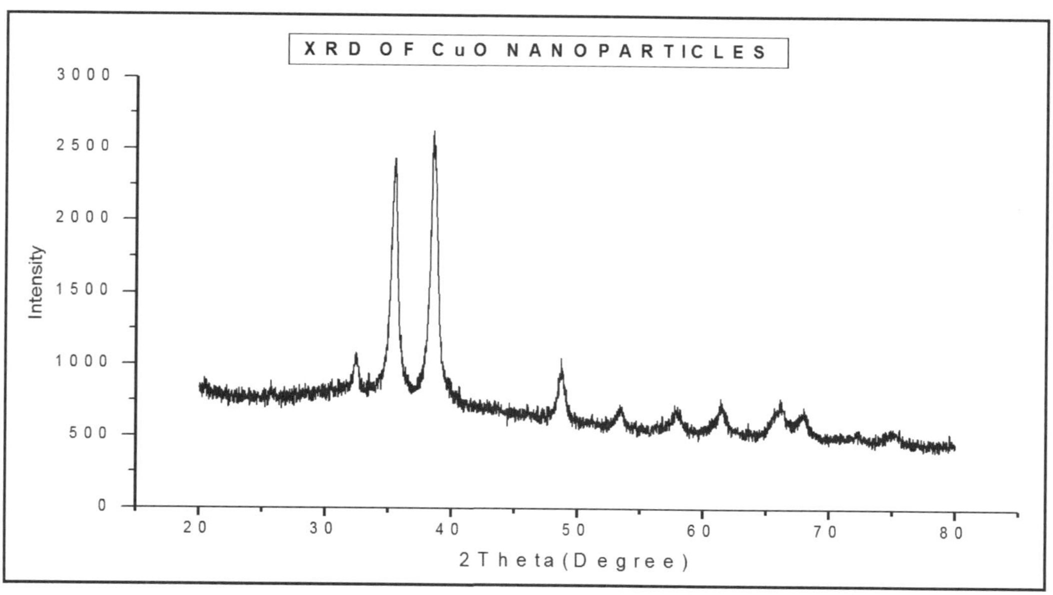

Fig. 2: X-Ray Diffraction Pattern of CuO Nanoparticles

Figure 3 exhibits the XRD pattern for PANi-CuO nanocomposites. The peak observed at $2\theta = 26.09°$ favors to pure PANi and at 32.9°, 35.7°, 40°, 50° and 53.5° to CuO. The remarkable shifting in the positions of 2θ values confirms the interactions between PANi chain and nanoparticles of CuO.

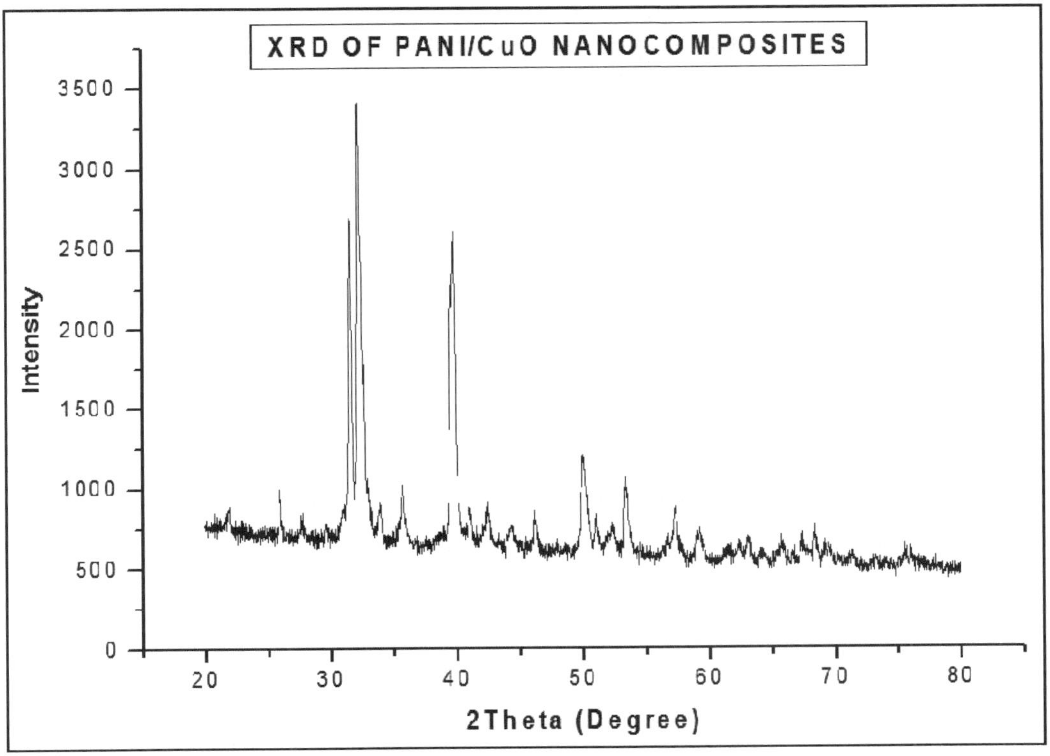

Fig. 3: X-Ray Diffraction Pattern of PANi-CuO Nanocomposite

The average particle size of prepared materials was calculated by Scherrer formula given by following equation,

$$D = \frac{0.9\lambda}{\beta \cos\theta}$$

Where,
D = Average crystallite size of the powdered particles,
λ = 1.54 Å (wavelength of CuKα),
β = Full Width at Half Maximum (FWHM) of intensity of major peak,
θ = Bragg angle (angular position of the peak) (Ambalagi *et al.*, 2018).

The crystallite size of the Cupric Oxide nanoparticles and pure PANi are found to be equal to 50.13 nm and 62 nm, respectively. The average crystallite size of the prepared nanocomposites is found to be 24.19 nm.

Fourier Transform Infrared Spectroscopy

The FTIR spectra of CuO nanoparticles and PANi-CuO nanocomposites are clearly exhibited in the Figures 4 and 5 respectively. In Figure 4 two vibrational bands at 668.35 cm^{-1} and 772.50 cm^{-1} and one stretching band at 1339.59 cm^{-1} attribute characteristic bands of Cu(II)O nanoparticles. One other band at 1558.51 cm^{-1} shows the Cu–O symmetrical stretching.

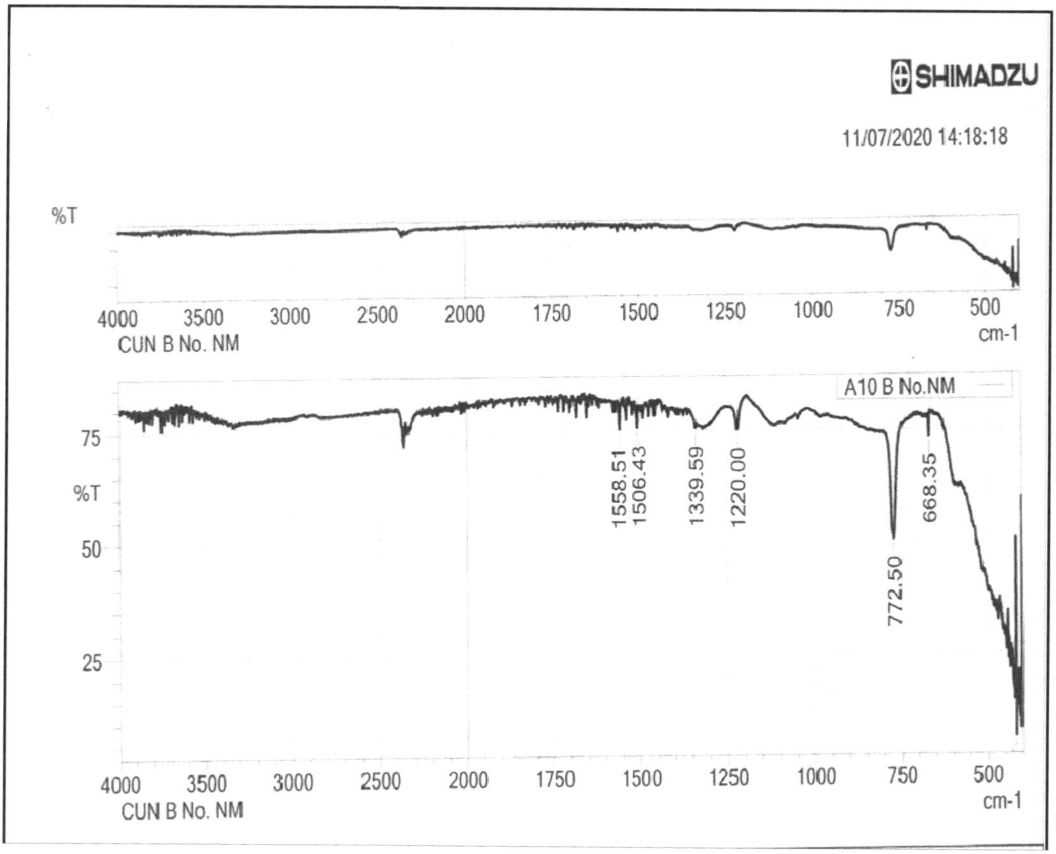

Fig. 4: FTIR Spectrum of CuO Nanoparticles

In contrast to pure PANi (Jundale *et al.*, 2013), the FTIR of PANi-CuO nanocomposites in Figure 5 represents some shifts in the wavenumbers. The reason for this shifting is the loss in conjugation after formation of PANi-CuO blend. The bands obtained at 1491.00 cm^{-1} and 1557.79 cm^{-1} exhibited the stretching mode of vibration for the benzenoid and quinonoid unit of PANi, respectively. The band at 1295.22 cm^{-1} attributes to C-H stretching. The region of bands 477.39-844.84 cm^{-1} confirms the presence of CuO particles in the nanocomposite.

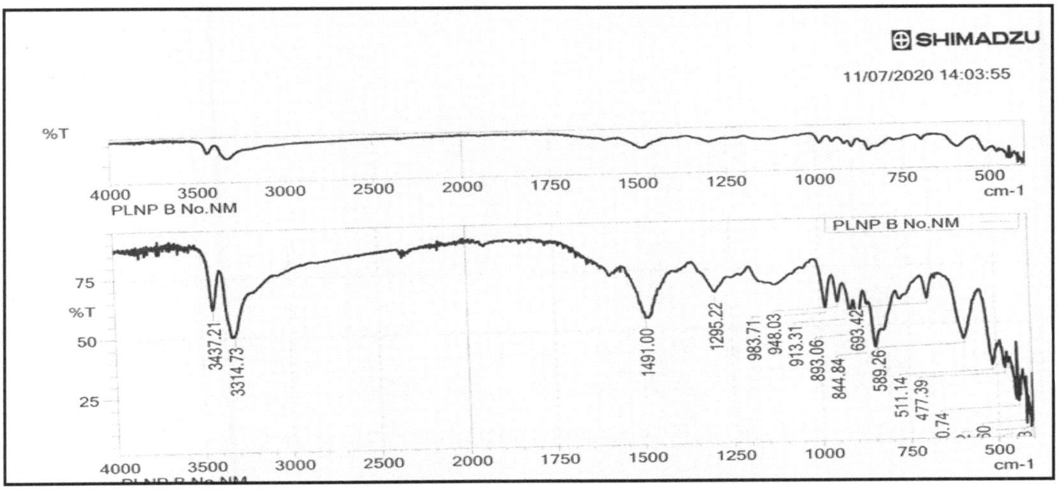

Fig. 5: FTIR Spectrum of PANi-CuO Nanocomposites

UV-VIS Spectrometry

Figures 6 and 7 indicates the UV-Visible spectra of CuO nanoparticles and PANi-CuO nanocomposite, respectively. The UV-Vis spectra of CuO nanoparticles in Figure 6 represents a strong absorption peak at 218 nm which confirms the formation of CuO nanoparticles as reported in earlier research (Dhineshbabu *et al.*, 2016). The UV-Vis Spectra of PANi-CuO nanocomposites in Figure 7 represents λ_{max} at 266.50 nm which is due to the π–π* transition of the benzenoid ring. The peak observed at 539.50 nm is due to the synergetic effect between Cupric Oxide and quinonoid ring of PANi (Athisayaraj, 2019).

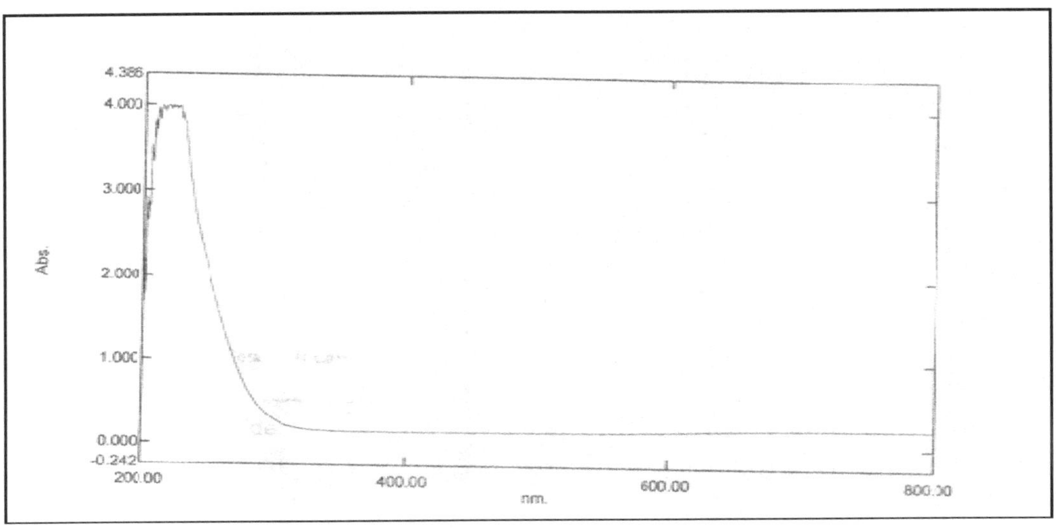

Fig. 6: UV-Vis Absorption Spectrum of CuO Nanoparticles

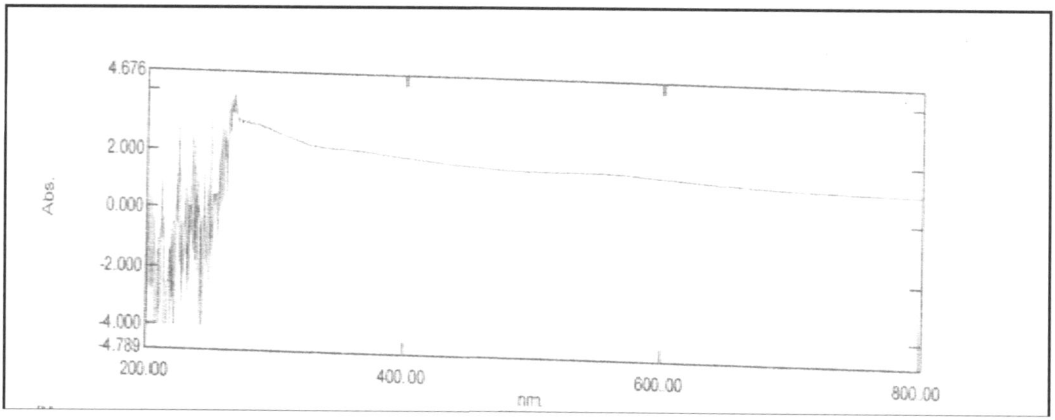

Fig. 7: UV-Vis Absorption Spectrum of Polyaniline–Cupric Oxide Nanocomposites

Scanning Electron Microscopy (SEM)

The surface morphology of CuO nanoparticles and PANi-CuO nanocomposite were represented in Figures 8 and 9, respectively. The SEM image of CuO nanoparticles in Figure 8 shows the fine, granular and spherical particles. The SEM image of composite in Figure 9 shows the strong effect on morphology of PANi on account of incorporation of CuO particles. It is observed that the crystallinity decreases in PANi-CuO nanocomposites when compared to CuO nanoparticles. This is due to the amorphous nature of PANi. Some smooth solid blocks are visible in the SEM image of PANi-CuO nanocomposites because of the presence of oxide particles, which increases the crystalline nature of synthesized materials.

Fig. 8: SEM Image of CuO Nanoparticles

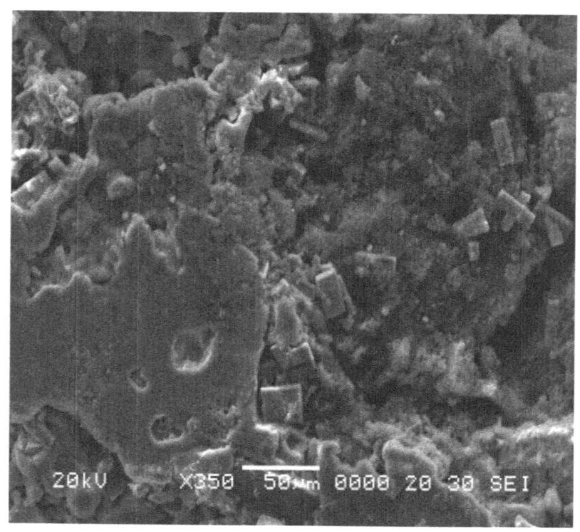

Fig. 9: SEM Image of PANi–CuO Nanocomposite

Electrical Studies

The electrical study is important property to explore various applications of synthesized materials in electrical devices. The DC conductivity of synthesized materials was measured by using Four-probe conductivity measurement. The conductivity of pure PANi and PANi-CuO nanocomposites were found to be equal to 1.6×10^4 S cm^{-1} and 3.1×10^4 S cm^{-1}, respectively. It is very clear from the above values that the conductivity increases when the pure PANI is converted into nanocomposites with CuO particles.

Antimicrobial Activity

The antimicrobial activity of the prepared nanocomposites was performed by Disk Diffusion method. The composite material was placed over the nutrient agar plates in concentrations of 5 mg/ml and 10 mg/ml. The antimicrobial activity of synthesized PANi-CuO nanocomposites against *Bacillus subtilis* (Gram-positive bacteria) and *Escherichia coli* (Gram-negative bacteria) is shown in Figures 10 and 11, respectively. The average zone of inhibition against pathogenic bacteria confirms the antibacterial property of synthesized conducting polymer.

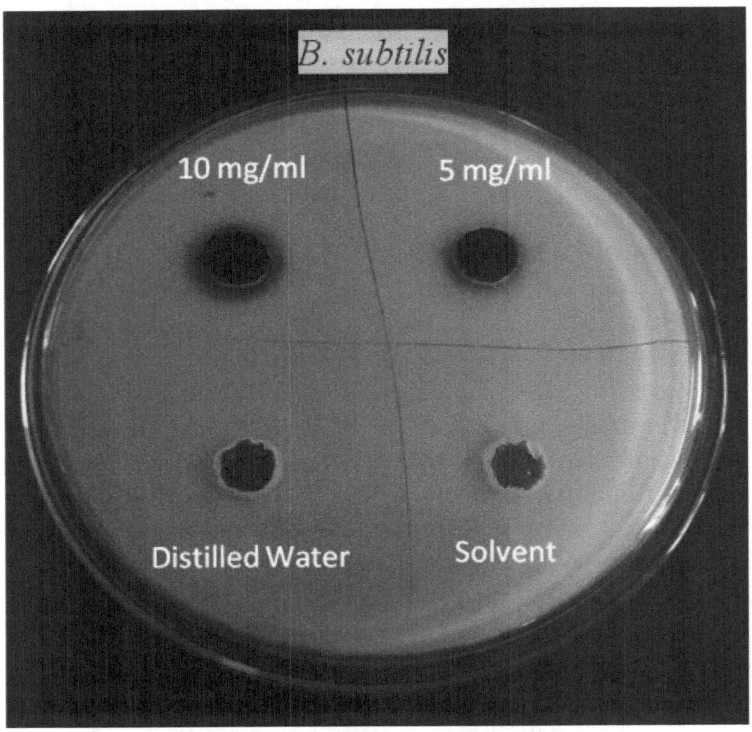

Fig. 10: Antimicrobial Activity of PANi Copper Oxide Nanocomposites against *B. subtilis*

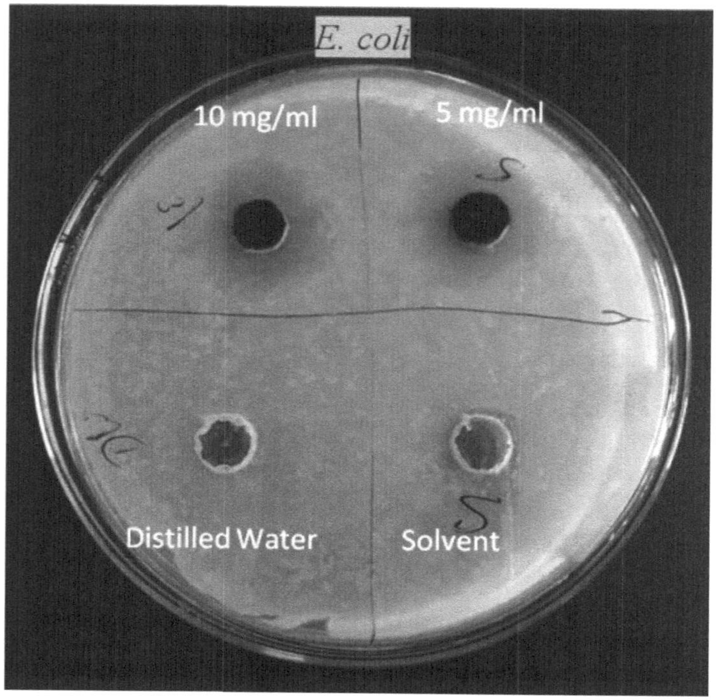

Fig. 11: Antimicrobial Activity of PANi Copper Oxide Nanocomposites against

CONCLUSION

The copper oxide nanoparticles were synthesized by chemical co-precipitation method and the pure PANi and PANi-CuO nanocomposites were successfully fabricated by chemical oxidative polymerization. The peaks obtained in XRD are in good agreement with the reported values. The average particle size of CuO, pure PANi and PANi-CuO nanocomposite have been calculated by Scherrer formula, which are found to be equal to 50.13 nm, 62 nm and 24.19 nm, respectively. The characteristic peaks of FTIR and UV-Vis spectroscopy and the SEM images confirms the successful synthesis of materials. From the Four-probe conductivity measurements, it is clear that the conductivity increases when the pure PANi is converted into nanocomposites with CuO particles. The antimicrobial results of the nanocomposites recommend that the synthesized nanocomposites exhibit microbicidal effect.

ACKNOWLEDGEMENT

We acknowledge Shri Vaishnav Vidyapeeth Vishwavidyalaya, Indore, for providing laboratory facilities. We are grateful to UGC-DAE Consortium for Scientific Research Center, Indore to provide analytical facilities.

REFERENCES

[1] Ambalagi, S.M., Devendrappa, M., Nagaraja, S. and Sannakki, B. (2018). Dielectric properties of PANI/CuO nanocomposites. IOP *Conference Series: Material Science and Engineering,* 310, 1–5.

[2] Ashokkumar, S.P., Vijeth, H., Yesappa, L., Niranjana, M., Vandana, M. and Devendrappa, M. (2020). Electrochemically synthesized PANi/copper oxide nano composites: To study optical band gap and electrochemical performance for energy storage devices. *Inorganic Chemistry Communications.,* 115, 107865-1-107865-11.

[3] Ashokkumar, S.P., Yesappa, L., Vijeth, H., Niranjana, M. and Devendrappa, H. (2017). Electrical conductivity and morphology of electrochemical synthesized PANi/CuO nano composites. *AIP Conference Proceedings,* 1953, 030222-1-030222-5.

[4] Athisayaraj, E.P.P. and Vedhi, C. (2019) Composites of copper oxide and PANi for solar cells applications. *International Journal of Advance Science, Research & Management,* 4, 23–28.

[5] Dhineshbabu, N.R., Rajendran, V., Nithyavathy, N. and Vetumperumal, R. (2016). Study of structural and optical properties of CuO nanoparticles. *Applied Nanoscience,* 6, 933–939.

[6] Elango, M., Deepa, M., Subramanian, R. and Musthafa, A.M. (2017). Synthesis, characterization and antibacterial activity of polyindole/Ag-CuO nanocomposites by reflux condensation method. *Polymer-Plastics Technology Engineering,* 57, 1440–1451.

[7] Ge, L., Han, C. and Liu, J. (2012). In situ synthesis and enhanced visible light photocatalytic activities of novel PANI–g-C_3N_4 composite photocatalysts. *Journal of Material Chemistry.,* 22, 11843–11850.

[8] Gulce, H., Eskizeybek, V., Haspulat, B., Sari, F., Gulce, A. and Avci, A. (2013). Preparation of a new PANi/CdO nanocomposite and investigation of its photocatalytic activity: comparative study under UV light and natural sunlight irradiation. *Industrial and Engineering Chemistry Research,* 52, 10924–10934.

[9] Jundale, D.M., Navale, S.T. and Khuspe, G.D. (2013). PANi–CuO hybrid nanocomposites: synthesis, structural, morphological, optical and electrical transport studies. *Journal of Material Science: Materials in Electronics,* 24, 3526–3535.

[10] Nadaf, L.I. and Venkatesh, K.S. (2015). PANi-copper oxide nano composites: synthesis and characterization. *Material Science Research India,* 12, 108–111.

[11] Shu'aibu, Y.I., Muhammad, U. & Muhammad, A.A. (2018). Humidity sensing study of PANi/copper oxide nanocomposites. *International Journal of Advanced Academic Research,* 4, 49.

[12] Singh, P. and Shukla, S.K. (2020). Structurally optimized CuO/PANi nanocomposites for efficient humidity sensing. *Surfaces and Interfaces,* 18, 100410-1 - 100410-13.

[13] Tamayo, L. and Azocar, M. (2016) Copper-polymer nanocomposites: An excellent and cost-effective biocide for use on antibacterial surfaces. *Material Science and Engineering C,* 69, 1391–1409.

[14] Vyas, S., Shivhare, S. and Shukla, A. (2017). PANi (PANI) metal oxide nanocomposites as a conducting material. *International Journal of Research and Scientific Innovation,* 4, 86–89.

[15] Zhou, Z., Zhang, X., Lu, C., Lan, L. and Yuan, G. (2014). PANi-decorated cellulose aerogel nanocomposite with strong interfacial adhesion and enhanced photocatalytic activity. *RSC Advances,* 4, 8966–8972.

An Effective Tiny Object Detection Approach Based on Deep Learning Methods

Sachin Chirgaiya[1] **and Anand Rajavat**[2]

ABSTRACT

A fascinating subject in computer vision is tiny object detection. With the exponential growth of deep learning, several researchers have entered the race. Area proposals, segregated grid cells, multiscale attribute maps, and a new loss mechanism are among the suggested developments. As a consequence, target detection efficiency has recently undergone substantial enhancements. The current state-of-the-art of many detectors, have failed to identify tiny objects in both single-stage and multi-stage methods. The latest state-of-the-art deep learning models, such as CNN, RCNN, Fast RCNN, Faster RCNN, RetinaNet, and YOLOv3, are examined in this study. However, a detailed evaluation of the benefits and weaknesses of models is made based on the available studies. Lastly, comparative observations and analyses are done and then discussed.

Keywords: Computer Vision, Enhancements, RetinaNet, Segregated Grid Cells, Target Detection.

INTRODUCTION

Tiny object detection is a work in which the bounding boxes and tagging of the objects are done. In this task the groups of objects are formed which they belong, and also the locations of distinct items are marked from an input image (Gong *et al.*, 2021). Several proposals have been identified and from them the conventional approaches are found to be better on the job in which deep learning is the base (Carion *et al.*, 2020). Tiny object detection approaches are primarily divided into two groups, called multi-stage approaches and regression (Duan *et al.*, 2020) or classification-based approaches (Samet *et al.*, 2020) or single-stage approaches (Zhao *et al.*, 2019) and are recognized as real-time and unified networks (Uzkent *et al.*, 2020). Because of the need for fulfilling everyday reality and helping people get a happier life, apps focused on real-time object detection

[1] Research Scholar, Shri Vaishnav Institute of Information Technology, Shri Vaishnav Vidyapeeth Vishwavidyalaya, Indore, Madhya Pradesh, India. E-mail: chirgaiya.sachin@svvv.edu.in

[2] Director, Shri Vaishnav Institute of Information Technology, Shri Vaishnav Vidyapeeth Vishwavidyalaya, Indore, Madhya Pradesh, India. E-mail: directorsviit@svvv.edu.in

have received a lot of attention these days (Rashwan *et al.*, 2020). Self-driving vehicles, for example, are an authentic one for helping people comfortably transport on the road at the same time, reducing traffic crashes by distracted drivers (Zhao *et al.*, 2019).

The other need to detect faulty component pieces in manufacturing industries, as well as the complexity of a view of angle, destructed shape and component size that dramatically varies during the installation process can also be addressed (Noh *et al.*, 2019). However, it shows that the detection of objects in a real-time application of world which is applied in computer vision, is very important. The goal of tiny objects detection, provided an image of interest, is to quickly determine what common objects are available in the image, particularly in very small sizes (Li *et al.*, 2017). According to different appearances of objects which are larger in size but due to change of view and length of image capturing object appears is small in size like bus, vehicle, traffic sign, person etc. or tiny objects like food item, cup, bottle, plate, etc. (Yao *et al.*, 2020), in Figure 1.

(a) Source: Yao *et al.*, 2020

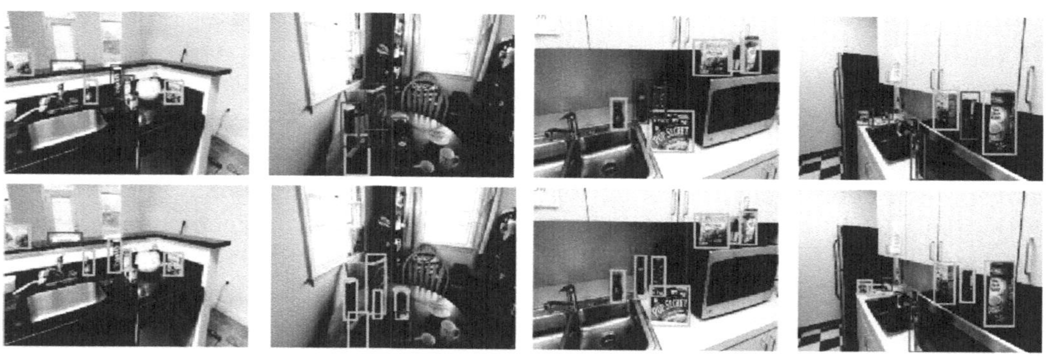

(b) Source: Liu *et al.*, 2020

Fig. 1: (a) Large Objects Occupying Very Small Appearance in an Image and (b) Actual Tiny Objects Appearing Tiny

Therefore, tiny object recognition in computer vision is a difficult task because small object representations and the large numbers of images are taken which makes this task more complex (Alahi *et al.*, 2016). For example, a picture is having various resolutions and sizes, when the resolution is low or high due to that it will stop to detect tiny objects through detector. During this process there may be some tiny objects which are deformed due to overlapping of the objects or may vary in size or shapes of the same objects. Ample amount of deep learning-based approach is proposed for the solution of this problem in recent years. Various proposals have been presented, with subsequent tests, to overcome object detection problems, but the existing detectors currently concentrate on detecting objects of typical sizes, not confining to tiny ones (Zhao *et al.*, 2019).

In the study of tiny object detection, an appraisal of tiny object detection methods is essential and significant. Object detection has received a lot of attention in the current research according to a current survey of the work based on the challenges taken from important databases like COCO and PASCAL VOC 2007 (Yao *et al.*, 2020). These datasets often include objects that take up medium to large sections on an image with a few tiny items, resulting in a data discrepancy between objects with various sizes and shapes, consequential in a preference of models against objects with more numbers. Furthermore, the amount of existing tiny object dataset classes is smaller than that of traditional datasets. Furthermore, most detectors have failed to identify tiny items, both in single-stage and multi-stage methods (Zhang *et al.*, 2018). The present paper provides an in-depth review of current deep learning frameworks to detect tiny items (He *et al.*, 2016).

With similar trade-off considerations, i.e., precision and execution time along with budget limitations, six state-of-the-art models have been tested, like CNN, R-CNN, Fast R-CNN, and Faster R-CNN of same family (Liu *et al.*, 2018), RetinaNet, and YOLOv3 too (Redmon & Farhadi, 2018). The present study summarizes benefits and weaknesses as well as model architecture rather than just implementing the concept. Hence, the inputs are as follows: (i) In two main detection approaches, called the single-stage approach and multi-stage approach, such as YOLOv3, RetinaNet, Fast R-CNN, and Faster R-CNN, along with common backbones such as FPN, ResNet, or ResNeXT (Zhao *et al.*, 2019), an extension to test deep models has been attempted, (ii) The benefits and weaknesses of models have been presented related to the accuracy, use of resources and processing speed of tiny items, as well as adjustments in these variables when the size of an object is scaled up or down, and also a distinction between single-stage and multi-stage strategies (Van Etten, 2018).

TINY OBJECT CHALLENGES

In this case many problems related to tiny object target recognition are addressed. Tiny item detection receives a lot of interest from researchers, but challenges still prevail; for example, COCO (Yao *et al.*, 2020) challenges include a standard for tiny object detection, but accuracy detection is still a challenge. TOD (Gong *et al.*, 2021), 1st Tiny Object Detection (TOD) Challenge which is a competition track: tiny person detection (Yu *et al.*, 2020). As a result, researchers have a difficult time identifying tiny objects because, in addition to the standard difficulties of object detection, tiny objects face special challenges. The concept of tiny objects is, however, not obviously simple as is discussed below.

Tiny Object Appearances

Tiny object detection has lately been regarded as an enticing issue in and of itself, with a number of problems that researchers find interesting. Due to small scale and size the contributors are still struggling to identify the objects without getting confused. Certain objects among many other objects that are positioned near or are of same size in appearance, the chances of the appearance of tiny items are much higher than other objects. It is really difficult to distinguish tiny items from the surrounding debris. Quiet contrary, the pixels available to represent tiny objects details are often much smaller than regular objects. In comparison, main features for obtaining tiny objects from an image are weak and also gradually lost as several kinds of various deep network layers such as pooling or convolutional layers go through. For example, in VGG16, if the object of interest has a size of 16 × 16, a limit of 1 pixel will be shown after 6 times passing through the pooling block. As a consequence, it is difficult to obtain successful results using tired scanning strategies like dramatic rise and sliding window in the number of bounding boxes (Gong *et al.*, 2021).

Tiny Object Definitions

The aim of this study is to identify the tiny object classification approach to find, how an object will appear in size and scale, as well as how many pixels it can contain in an image. However, the object may appear different due to high-resolution or low-resolution as the image plays an important role for tiny object detection. According to this example, an object is allocated as a tiny object to occupy a part of the resolution of 500 × 500 on 2048 × 2048 but to be very large on one of 800 × 800. When the dataset has the images with different resolution and aspect ratios, it causes problem for researchers. There are several descriptions of tiny objects that have proposed but they are not well defined. It relies on databases that are used for objects of interest for assessment and characteristics (Gong *et al.*, 2021).

Collection of Datasets

There are far more works that focus on various tiny items, resulting in a lack of expertise and knowledge to perform a detailed investigation. The previous methods clearly state that the emphasis has been on large objects and that the presence of tiny items has been overlooked. The present study test the efficiency of current state-of-the-art detectors in order to get a general image of their ability to detect tiny items. There are only a few works on the issue of tracking tiny objects in terms of tiny object detection. Most of these works have so far been developed specifically to classify such types, such pedestrians, vehicles, or as traffic signs which in the real world do not contain general or multi-class datasets (Zhu *et al.*, 2016). This results in a lack of appraisal of the tools to illustrate their capacity to detect multiple types of objects and even to alter their shapes (Li *et al.*, 2017). Microsoft COCO and Tiny Object datasets have been created and contain typical items like "cup," "telephone," "plate," "remote," "clock," "tissue box," "bottle," "pan," and "jar" (Bai *et al.*, 2018).

Microsoft COCO with PASCAL VOC 2007 is in general consisting of tiny items in different categories. Diverse sizes of objects influence detection efficiency, and therefore familiar meanings of tiny items to filter objects that match these definitions and form a dataset of four subsets corresponding to four different definitions of tiny items. In addition, in a task called Perception Meets Drones, there is recently a small object dataset: A Challenge (https://motchallenge.net), and this dataset is considered a daunting dataset since it consists of many tiny items, including tiny objects in photographs in diverse situations and wild environments, but the image views are snapshots of drones travelling overhead and taking pictures from images. However, this dataset does not have research annotations, so it is difficult to take it for assessment (Duan *et al.*, 2020).

OBJECT DETECTION MODELS

In recent years, convolution neural network (CNN) methods have seen many advances in widespread innovations in deep learning and have obtained strong results in diverse tasks. As a consequence, it is widely used to describe present works effectively. The majority of the works have demonstrated substantial advances in the detection of objects that fill an image with medium or large sections. The following approaches are used to boost R-CNN forms as pioneers. In Faster R-CNN solution is seen as a state-of-the-art approach. Though this collection of complex works employs a number of innovative and revolutionary ideas ranging from sliding windows to object proposals, and produces the best results as state-of-the-art methods on difficult datasets like MS-COCO and PASCAL VOC 2007, their implementations is time-consuming as it operates on an image which may result in a reduction in the detector's running efficiency.

As a result, despite reaching high precision, the detectors face trouble using them for real-time object detection. This means they only care about precision and disregard the consequences of processing speed. Furthermore, detecting tiny items in the physical world is just as critical as detecting large or medium-sized objects, if not more so. In particular, data must be processed promptly and correctly in the automobile, smart vehicle, military and smart transportation sectors to ensure that safety comes first. However, in most of these instances, the data recorded is far from our location, and the information is negligible.

Region Based Convolutional Neural Networks (R-CNN)

Region Based Convolutional Neural Networks (R-CNN) is an innovative concise advance approach that outperforms previous works on PASCAL VOC 2007 with 35% more mean average precision (mAP). The R-CNN design summary consists of four major steps that are known as this method's latest progress. R-CNN starts by resizing a picture to 228 × 228 as an input and selective search algorithm is applied over the input data and produces 3000 candidates as the twisted regions used for the feedback of the CNN function network of suggested bounding boxes. The computation of features is done based on network derives through a 4096-dimensional function as it moves through the regions. Finally, behind the last sheet, a class-specific linear SVM classifier is used to identify regions in order to decide whether there are any objects and what they are.

The value of features is vital to the R-growth. In CNN the HOG based low-level characteristics of images are taken and replaced with the CNN characteristics in R-CNN, which are potentially more discriminatory showoff. The assessment of a picture, however, is incredibly expensive and inefficient because the convolutional network must be implemented 2000 times by R-CNN. Furthermore, resizing the input to a low resolution of 228 × 228 is an issue that affects tiny items, which are quickly deformed or losing information when the resolution is adjusted faraway to its original measurements. The regional proposals contributed several times for computation of features, which must be placed on disks for each area proposal before conducting feature extraction. In addition, several overlapping bounding boxes can result in a decrease of mAP where tiny objects are similar to large objects since the bounding boxes holding large objects and indifference of bounding boxes for tiny objects are skewed to select (Zhao *et al.*, 2019).

Fast Region Based Convolutional Neural Networks (Fast R-CNN)

Fast R-CNN is an innovative approach which uses the deep convolutional networks to minimize training and testing time effectively to identify object proposals increasing accuracy. Multitask failure is done in Fast R-CNN through end-to-end training. Many RoIs are identified in an image with any image size taken through convolutional network. Many convolutional layers of the base network are use as a function map in the Fast R-CNN to add different RoIs, rather than adding wrapping them to feed into the network as input to Fast R-CNN does. A pooling layer extracts each RoI as a fixed-size feature vector, which is then translated to a feature vector by completely linked layers. The network has two performance vectors per RoI: probabilities of softmax and offsets of bounding-box regression per-class. Though forward and backward passes from the same picture to share computation and memory is the most critical aspect of RoI enabler in it.

In a Fast R-CNN a new way of training of the addresses is proposed with the limitations over come from R-CNN by increasing the accuracy rate and running time. Over all the mean detection accuracy is to find better as compared to R-CNN. Based on the multitask failure; the training process is a single phase which will refresh the entire network layers in a Fast R-CNN (Chen *et al.*, 2020).

Faster Region Based Convolutional Neural Networks (Faster R-CNN)

Over the Fast R-CNN, Faster R-CNN is a groundbreaking solution approach. It uses its region proposal network (RPN) as a tool instead external algorithms of generating bounding boxes which is trained end-to-end to produce highly skilled region proposals. After obtaining deep features from early convolutional layers the RPN is taken into account and windows slide over is considered for the feature map to remove features for each area proposal. The predications of the object bounding boxes are completely based on convolutional network in a RPN. The input image of any size is taken and a series of bounding boxes are obtained as an output over the images in a RPN through rectangular object recommendations with object ranking. In this task the over the fifth convolutional layer (conv5) the RPN takes a input image function by adding the 3×3 sliding windows to it.

The hidden layers as fed into two branches with one for object score deciding for the items and other for regression to decide the bounding box with similar ground truth. Since the RPN decreases the expense by sharing computation on convolutional features, it increases precision and running time whilst also 9capabilities, RPN and Fast R-CNN are integrated into a single network and allow Faster R-CNN to achieve high precision and efficiency but due to multistage of work the processing speed is suffered (Chen *et al.*, 2020).

Spatial Pyramid Pooling (SPP)

The SPP is design based on the CNN as a backbone which receives input with the fixed size of image like 224×224 so the raw images are frequently cropping from the original image truncation is done for the practical use of wrapping RoI (Lin *et al.*, 2017). For the arbitrary input size of the images a fixed-length of input is passed to convolution layer and it works as bridge to mediate and fully connect layer in a SPP. The SPP-network, in particular, uses the R-CNN method to find 3000 candidates for area proposals before extracting the function maps from the entire image. However, in SPP maps window of the features correspond to RoI with fixed length. Finally, SVM is used to distinguish using two completely linked layers and the R-CNN vs. SPP, the SPP has 100 times faster detection (Zhao *et al.*, 2019).

Single Shot MultiBox Detector (SSD)

Deep neural based single-stage real-time object detector is formed to detect the multiple objects from an image and is called as Single Shot MultiBox Detector (SSD)(Xiang *et al.*, 2018). Faster R-CNN, from the state-of-the-art multi-stage processing technique is an object proposal and uses those to identify objects and objectness instead of using an external system for real-time detection. By removing the need for the proposed network, the SSD increases the runtime speed better than the previous detectors. As a result, there is a small decrease in mAP, which SSD compensates for by adding certain enhancements such as multiscale functionality. These upgrades allow SSD to use lower resolution images to gain the same Faster RCNN, which then further speeds up SSD processing (Liu *et al.*, 2016).

You Only Look Once (YOLO)

For multiple categories the real-time state-of-the-art object detection shows that You Only Look Once (YOLO) is to have the better results from the detector families. Currently YOLO has 3 versions which are expanded significantly with each update (Redmon & Farhadi, 2018). The following $A = \pi r^2$ detailed reviews of the YOLO techniques as a basis for bringing them into practice: YOLO v1 is a fully novel method which uses the single-stage network for object detection in a real time (Redmon & Farhadi, 2018). The core concept for YOLO detection is to divide images into grid views for object localization at real time through which the precision and running time improves. The aim of YOLO is to solve two issues: to identify the object and its location in an image. It observes continuously, quickly and conveniently with three key measures. To begin, YOLO first it resizes an input image with fixed size then passes it to single convolutional network. Finally uses the model's confidence score to set a threshold on the detections. The bounding boxes and confidence predication is directly depending on each grid cell that demonstrates how sure an individual is in the bounding box model as well as how precise it means that the bounding box is expected. YOLO's downside is that it lags behind the state-of-the-art precision detector technologies, but is better than those for operating time. As compared to Fast R-CNN, it generates less than half the amount of context errors.

YOLO v2 over comes the drawbacks of YOLO v1 with its updated features (Redmon & Farhadi, 2018). Similar to the original one the YOLO v2 works on various fixed sized input images with added detection and classification technique. YOLO v2 can train on multiclass datasets such as ImageNet or COCO and these are the novel enhancements. Furthermore, the detector was trained to identify over 9000 different object types. Customized network architecture is designed for the internal work of the YOLO v2. According to the current state-of-the-art detectors, YOLO v2 focuses specifically to enhance precision and localization while receiving high classification precision. Despite the fact that YOLO v2 increases accuracy, but it fails to detect tiny objects due to the input down sampling, as a result the final forecast due to low dimension function mapping results are not up to the mark.

YOLO v3 comes with great enhancements towards the tiny object detection, in order to solve major issues of the YOLO v2 (Redmon & Farhadi, 2018). In general, with deep characteristics acquired from multiple layers, a number of recent networks appear to be stronger and deliver good results on their tasks. The way of computing the cost function is also updated by YOLO v3. The resulting objectness score should be 1 if the anchor overlaps a ground truth rather than all bounding boxes. Other anchor boxes that have overlapped greater than a predefined level of 0.5 are eligible. Only one boundary box is connected with each ground fact. There is no lack of classification or localization if a bounding box is not allocated, just a loss of faith in objectness.

Drawbacks of CNN

According to state-of-the-art networks several CNN models are based on different layers like pooling and convolutional. However, travelling through several levels of detail is not a reasonable way to find tiny objects, since the objects of interest in tiny object detection is a very complex task. Due to small appearances and little in size the object detection is very expensive. Furthermore, tiny objects are highly vulnerable to shifts in image sizes, unlike regular or massive. In CNN the images move from layer to layer due to that some useful features are removed and the sizes of the images are reduced.

Even, there won't be much to note if tiny particles only go through convolutionary layers. Tiny objects with only a few informative properties must move through convolutional and pooling layers, which assist in preventing overfitting of data and lowering evaluation costs by lowering number of parameters. For the training process, we will take advantage of a manner in which the method produces information to address the data limitations of tiny items. While images do need to transfer layers such as pooling and convolutional layers, the network just has less layers relative to those in this sense.

Experimental Setting

On three datasets, MS-COCO, PASCAL VOC 2007 and a small object dataset, are continuously trained and tested for different object detectors (S. Liu *et al.*, 2019). During this time, the approaches evaluated consist of RetinaNet, YOLOv3, and Faster R-CNN, over the various backbones. PASCAL VOC 2007 and COCO are the original datasets that are widely used in target detection today (Deng *et al.*, 2020). Both databases are made up of virtually huge objects and other types of objects with variable size, whose size makes up a large part of the image. For tiny object detection, these two datasets are not sufficient (Chen *et al.*, 2020). In addition to that, the VisDrone dataset is a large-scale dataset that contains a large range of classes for tiny object detection obtained by drones. Nevertheless, it does not publish the test series labels for review, and the picture views are top-down, which is not the existing situation (Stojnić *et al.*, 2021). This dataset is a mixture of COCO and SUN datasets which are considered as tiny object dataset.

The Cup, Plate, Telephone, Remote, Clock, Bottle, Pan, Jar, Tissue box are among the following classes in the tiny object dataset. There are 4925 images in total in the dataset, with 3296 images for preparation and 1629 images for research. Table 1 lists the specifics of the number of tiny objects and the images they hold for the dataset sub-sets (Zhu *et al.*, 2018). All models were trained and tested on tiny object datasets with the same parameters. In order to train and test a particular model with 70k iterations were trained in the training process with parameters including training time, batch size, learning rate, and stage size. An attempt was made to lunch the model with 100k iterations and it was observed that the data was lost with higher learning rate.

Table 1: List of Classes and Subsets

Subsets	*Classes*	*Images*	*Objects*
PASCAL VOC_5	15	329	529
PASCAL VOC_10	20	2231	5893
PASCAL VOC_20	18	2970	7867
PASCAL VOC_25	20	1070	2313

An attempt was to converge at a lower learning rate with trial to understand whether the models would act as a starting point. Nevertheless, everything remained unchanged. Also, an exponential growth was observed with first iterations over 10 K and then steadily slowing down after 20 K. As a result, training and learning rate was eventually reduced by 25 K. With the following setting the loss value gets fixed after 40 K iterations and after that the training rate increased with 70 K to see the changes in loss values. It was observed that there is no change after 40 K iterations. The attempt to test the 30 K to 70 K models, and after 40 K iterations, the performance of the models was generally not consistent. For this cause, at 30 K and 40 K iterations, the weight for measurement was gathered. YOLO gets the better performance after 30 K iterations, while others get the best after 40 K iterations.

RESULTS AND ANALYSIS

This section covers the findings from the experimental process and testing of the existing models. All models in this section were trained on the same environment and GPU: Ubuntu 18.04.4 LTS, CPU @ 3.10 GHz, GPU Tesla P100 with 16 GB RAM. Other comparisons were made in addition to the comparative precision to ensure that the evaluation findings of the present study are objective and consistent.

Testing Accuracy of the Models

According to Table 2 the multi-stage approaches performed well over the single-stage approaches with the accuracy range around 8–10%. The multi-stage approaches with Faster R-CNN along with the FPN backbone and RestNeXT with same backbone obtained the highest mAP. YOLO v3 with Darknet-53 scored 48.87% higher than the top in single-stage approaches. Currently, in the sense of tiny object dataset, this is also true. With the slightly upgrade of Fast RCNN with same backbones the Faster RCNN is performed well with just 1–2% stronger than Fast RCNN for ResNeXT backbones and is similar to Fast RCNN for the rest. The discrepancy isn't important, because it is specific to the situation. Depending on the backbone, Fast RCNN achieves accuracy ranging from 16.7% to 39.05%. Faster RCNN, on the other hand, earns 30.1% to 41.2%. Second, YOLO outperforms SSD and RetinaNet in one-stage approaches.

Table 2: Observation of Sub-Class Objects

Method	*Backbone*	*Cup*	*Plate*	*Tel.*	*Remote*	*Clock*	*Bottle*	*Pan*	*Jar*	*T. box*	*mAP*
YOLO	Darknet-19	21.95	37.85	4.65	60.75	28.65	12.45	18.75	9.15	36.35	23.52
		29.35	37.65	3.75	61.15	33.65	13.85	23.95	10.05	27.45	25.01
	Darknet-53	30.22	48.8	12.28	61.44	44	16.62	**31.38**	16.1	49.05	**31.27**
		31.73	49.54	12.21	66.83	**48.87**	18.84	31.87	15.57	47.31	34.09
	ResNet-50	25.53	38.3	1.85	55.18	19.22	14.5	20.43	10.79	36.43	23.18
		28.67	42.3	3.22	62.81	30.03	15.79	20.89	11.04	36.76	26.49
	ResNet-101	27.47	37.25	4.48	56.68	23.25	16.36	17.91	10.27	39.46	23.63
		30.54	46.24	10.87	63.13	34.19	16.28	24.53	14.19	39.92	29.59
	ResNet-152	27.45	38.19	2.34	54.76	21.45	14.88	20.86	12.44	33.81	22.66

Method	*Backbone*	*Cup*	*Plate*	*Tel.*	*Remote*	*Clock*	*Bottle*	*Pan*	*Jar*	*T. box*	*mAP*
		27.76	46.19	6	61.95	32.61	13.77	23.32	13.3	40.74	27.57
SSD	ResNet-101	7.25	10.75	1.45	26.45	6.95	5.25	0.65	5.45	18.95	9.24
	VGG16	10.85	18.75	1.45	27.05	9.95	9.85	0.65	5.45	17.45	10.15
	VGG16	10.85	18.75	1.45	43.95	9.95	9.85	9.75	10.05	8.35	12.31
RetinaNet	ResNet-50-FPN	32.45	50.95	3.45	66.45	22.15	16.85	9.15	13.85	26.45	24.29
	ResNet-101-FPN	32.35	50.35	8.55	65.65	20.85	16.65	12.45	11.65	39.45	26.09
	ResNeXT-101-32 8d-FPN	**37.25**	**56.65**	**13.55**	**67.45**	24.75	**19.15**	10.45	**17.15**	**54.45**	30.99
R-CNN	RPN prop. + VGG16	33.65	32.95	5.65	57.75	31.95	10.05	14.85	17.35	30.15	25.79
	Alexnet, 7, 300 pro	34.15	28.85	6.55	57.85	28.85	10.55	14.25	13.35	36.35	24.89
	VGG16, 7, 300 pro	39.05	31.95	8.65	61.55	42.35	16.55	22.15	14.65	34.05	29.39
	ContextNet (Alexnet, 7)	34.45	28.45	6.05	57.35	27.15	10.65	13.55	13.15	34.75	24.49
Fast RCNN	ResNet-50-C4	34.15	47.95	7.95	66.75	39.15	20.85	25.95	17.55	52.75	32.69
	ResNet-50-FPN	39.15	48.95	8.75	69.85	47.55	21.75	32.75	18.05	46.65	34.29
	ResNet-101-FPN	41.05	51.95	12.05	69.25	47.95	21.15	33.95	19.55	52.15	36.49
	ResNeXT-101-32 8d-FPN	49.25	56.45	11.75	72.75	54.85	22.15	35.05	22.65	54.25	39.69
Faster R-CNN	VGG16	25.51	39.3	9.48	54.95	17.01	12.63	15.77	10.05	38.04	22.91
	ResNet-50-FPN	37.45	51.55	8.75	69.35	49.75	19.55	30.25	15.65	54.05	34.79
	ResNet-101-FPN	41.55	50.85	6.35	69.15	47.85	19.25	30.35	14.95	52.95	34.69
	ResNeXT-101-32 8d-FPN	*51.55*	58.25	12.85	73.05	*57.15*	23.95	*37.65*	*21.75*	59.45	41.49

YOLO, on the other hand, earns the highest score of 34.1 points, while SSD and RetinaNet receive 10.15%, respectively. YOLO and SSD are called cutting-edge approaches in terms of speed without losing precision. In single-stage methods, RetinaNet is allocated to a process, but it cannot operate in real time. However, if the resolution is too poor in contrast to the original image format, precision suffers. For example, the precision of YOLO v3 with Darknet-19 is lower than the resolution of 680 × 680. In addition, increasing Darknet-53's resolution was also tried from 416 to 680, but the mAP declined when the resolution reached 680 × 680. As a consequence, for models like SSD and YOLO, the influence of picture size is apparent. In general, all mAP comparative findings on this dataset display a significant number of groups dominating, which is due to a data imbalance between the images and instances.

Trade-Off among Detectors

The trade-off is also impacted by decision, which was adjusted during model training and research. The best resolution is to be selected to ensure that the models work properly. In the case of multi-stage methods, the concept of using area proposals to boost object localization for detection is also a successful one. This is helpful, but it should be noted that ideas can be created on function maps rather than directly on input images because this has a huge effect on how models operate and how representations of objects is classified. It is fine for models to work if objects are with normal or medium appearance, but if objects are in multiclass, this is a challenge to consider and study extensively to match and enhance efficiency. Over all, as the size of objects is adjusted, all the models to be tested need to be changed, and the precision of the models varies a lot, with the exception of Faster RCNN.

While the accuracy of VGG16 is not superior to that of other architectures, it differs in that as it does not vary greatly in accuracy. Which is only acceptable for large items with a bounding box and image overlap of more than 10%; otherwise, this is not ensured. Figure 1 indicates that tiny objects have more options than bigger objects. The bus black length resembles a car on area in several respects. This probability of small object presence helps detectors operate harder and leads to inaccurate detection. Tiny objects may appear anywhere in a picture, resulting in the fact that detectors often create incorrect detections of recognizable appearances.

CONCLUSION

Tiny object detection is a difficult and fascinating task of object detection that has gained researchers' attention now a days. The advancement in deep learning has encouraged to enhance task efficiency in computer vision. However, the deep detection models were developed to solve problems related to general object detection, they also contribute to the performance of tiny object detection on a specific level. Despite recent achievements, detection efficiency has increased considerably, and there is still a substantial variation in precision between normal and tiny objects. The disparity between small, medium, and large scales is too great in the COCO dataset's parameters. Most models are effective at detecting normal objects, but when used to detect tiny objects, problems can be encountered. As a result, the first step in narrowing the gap in tiny object detection is to invest in datasets with vast quantities of data to train models on, as well as a wide variety of categories to compete with the human visual system.

REFERENCES

[1] Alahi, A., Goel, K., Ramanathan, V., Robicquet, A., Fei-Fei, L. and Savarese, S. (2016). Social lstm: Human trajectory prediction in crowded spaces. *Proceedings of the IEEE Conference on Computer Vision and Pattern Recognition*, 961–971.

[2] Bai, Y., Zhang, Y., Ding, M. and Ghanem, B. (2018). Sod-mtgan: Small object detection via multi-task generative adversarial network. *Proceedings of the European Conference on Computer Vision (ECCV)*, 206–221.

[3] Carion, N., Massa, F., Synnaeve, G., Usunier, N., Kirillov, A. and Zagoruyko, S. (2020). End-to-end object detection with transformers. *European Conference on Computer Vision*, 213–229.

[4] Chen, Y., Zhang, Z., Cao, Y., Wang, L., Lin, S. and Hu, H. (2020). Reppoints v2: Verification meets regression for object detection. *ArXiv Preprint ArXiv:2007.08508.*

[5] Deng, C., Wang, M., Liu, L. and Liu, Y. (2020). Extended feature pyramid network for small object detection. *ArXiv Preprint ArXiv:2003.07021.*

[6] Duan, K., Xie, L., Qi, H., Bai, S., Huang, Q. and Tian, Q. (2020). Corner proposal network for anchor-free, two-stage object detection. *ArXiv Preprint ArXiv:2007.13816.*

[7] Gong, Y., Yu, X., Ding, Y., Peng, X., Zhao, J. and Han, Z. (2021). Effective Fusion Factor in FPN for Tiny Object Detection. *Proceedings of the IEEE/CVF Winter Conference on Applications of Computer Vision*, 1160–1168.

[8] He, K., Zhang, X., Ren, S. and Sun, J. (2016). Deep residual learning for image recognition. *Proceedings of the IEEE Conference on Computer Vision and Pattern Recognition*, 770–778.

[9] Li, J., Liang, X., Wei, Y., Xu, T., Feng, J. and Yan, S. (2017). Perceptual generative adversarial networks for small object detection. *Proceedings of the IEEE Conference on Computer Vision and Pattern Recognition*, 1222–1230.

[10] Lin, T.-Y., Dollár, P., Girshick, R., He, K., Hariharan, B. and Belongie, S. (2017). Feature pyramid networks for object detection. *Proceedings of the IEEE Conference on Computer Vision and Pattern Recognition*, 2117–2125.

[11] Liu, L., Ouyang, W., Wang, X., Fieguth, P., Chen, J. and Liu, X. (2018). Deep Learning for Generic Object Detection. *A Survey [J].*

[12] Liu, S., Huang, D. and Wang, Y. (2019). Learning spatial fusion for single-shot object detection. *ArXiv Preprint ArXiv:1911.09516.*

[13] Liu, W., Anguelov, D., Erhan, D., Szegedy, C., Reed, S., Fu, C.-Y. and Berg, A. C. (2016). Ssd: Single shot multibox detector. *European Conference on Computer Vision*, 21–37.

[14] Liu, Z., Gao, G., Sun, L. and Fang, Z. (2020). HRDNet: high-resolution detection network for small objects. *ArXiv Preprint ArXiv:2006.07607.*

[15] Noh, J., Bae, W., Lee, W., Seo, J. and Kim, G. (2019). Better to follow, follow to be better: Towards precise supervision of feature super-resolution for small object detection. *Proceedings of the IEEE/CVF International Conference on Computer Vision*, 9725–9734.

[16] Rashwan, A., Agarwal, R., Kalra, A. and Poupart, P. (2020). Matrixnets: A new scale and aspect ratio aware architecture for object detection. *ArXiv Preprint ArXiv:2001.03194.*

[17] Redmon, J. and Farhadi, A. (2018). Yolov3: An incremental improvement. *ArXiv Preprint ArXiv:1804.02767.*

[18] Samet, N., Hicsonmez, S. and Akbas, E. (2020). HoughNet: Integrating near and long-range evidence for bottom-up object detection. *European Conference on Computer Vision*, 406–423.

[19] Stojnić, V., Risojević, V., Muštra, M., Jovanović, V., Filipi, J., Kezić, N. and Babić, Z. (2021). A Method for Detection of Small Moving Objects in UAV Videos. *Remote Sensing*, *13*(4), 653.

[20] Uzkent, B., Yeh, C. and Ermon, S. (2020). Efficient object detection in large images using deep reinforcement learning. *Proceedings of the IEEE/CVF Winter Conference on Applications of Computer Vision*, 1824–1833.

[21] Van Etten, A. (2018). You only look twice: Rapid multi-scale object detection in satellite imagery. *ArXiv Preprint ArXiv:1805.09512.*

[22] Xiang, W., Zhang, D.-Q., Yu, H. and Athitsos, V. (2018). Context-aware single-shot detector. *2018 IEEE Winter Conference on Applications of Computer Vision (WACV)*, 1784–1793.

[23] Yao, Y., Wang, Y., Guo, Y., Lin, J., Qin, H. and Yan, J. (2020). Cross-dataset training for class increasing object detection. *ArXiv Preprint ArXiv:2001.04621.*

[24] Yu, X., Gong, Y., Jiang, N., Ye, Q. and Han, Z. (2020). Scale match for tiny person detection. *Proceedings of the IEEE/CVF Winter Conference on Applications of Computer Vision*, 1257–1265.

[25] Zhang, S., Wen, L., Bian, X., Lei, Z. and Li, S. Z. (2018). Single-shot refinement neural network for object detection. *Proceedings of the IEEE Conference on Computer Vision and Pattern Recognition*, 4203–4212.

[26] Zhao, M., Cheng, L., Yang, X., Feng, P., Liu, L. and Wu, N. (2019). TBC-Net: A real-time detector for infrared small target detection using semantic constraint. *ArXiv Preprint ArXiv:2001.05852.*

[27] Zhao, Q., Sheng, T., Wang, Y., Tang, Z., Chen, Y., Cai, L. and Ling, H. (2019). M2det: A single-shot object detector based on multi-level feature pyramid network. *Proceedings of the AAAI Conference on Artificial Intelligence*, *33*(01), 9259–9266.

[28] Zhao, Z.-Q., Zheng, P., Xu, S. and Wu, X. (2019). Object detection with deep learning: A review. *IEEE Transactions on Neural Networks and Learning Systems*, *30*(11), 3212–3232.

[29] Zhu, P., Wen, L., Bian, X., Ling, H. and Hu, Q. (2018). Vision meets drones: A challenge. *ArXiv Preprint ArXiv:1804.07437.*

[30] Zhu, Z., Liang, D., Zhang, S., Huang, X., Li, B. and Hu, S. (2016). Traffic-sign detection and classification in the wild. *Proceedings of the IEEE Conference on Computer Vision and Pattern Recognition*, 2110–2118.

Characterization of Pen Ink using Hyperspectral Imaging (VSC-6000)

Harendrakumar Bamburde[1] **and M.P. Goutam**[2]

ABSTRACT

Forensic analysis of ink plays a significant role in questioned document examination. The composition of ink is commonly kept confidential by the manufacturers and varies from brand to brand. It is possible to differentiate the handwriting written with different brands of ink with hyper spectral imaging (VSC-6000) to individualization of ball point pen ink of different manufacturers. Different types of documents were selected, collected from different sources. These samples were altered by color inks belonging to different companies and were examined under Visual Spectral Comparator for ink differentiations under filtered light within the range 400 to 1000 nm. Study revealed that VSC analysis is suitable and reliable method for the differentiation of ink brands.

Keywords: Hyperspectral Imaging, Ink Differentiations, Questioned Document, VSC-6000.

INTRODUCTION

The forensic examination of writing inks on questioned documents has been performed since the beginning of 20th century. The inks contain different types of substances and the most important component among them are the pigments. There are some environmental factors like UV light that can cause ageing of inks moreover, due to chemical changes composition may also change. Forensic document examiners usually face variations/diversity in the ink formulations moreover, in brands and/or models of the writing instrument, specially when source of ink under investigation is to be determined. Therefore, analysis of ink is essential to reveal the secret clues of the questioned document. Forensic analysis of pen inks that may be either for identification, comparison or dating purposes, it remains essential part of the cases involving suspected document fraud (Neumann, 2009; Siegel, 2013). Forensic document examiners nowadays are using VHC workstations for specialized lighting inspections (Koppenhaver, 2007; Lewis, 2014). Reed *et al.* (2014) advocated the utility and advantages of Hyper

[1] Assistant Professor, Department of Forensic Science, Government Institute of Forensic Science, R.T. Road Civil Lines, Nagpur, Maharashtra, India. E-mail: hkcb7fodex@gmail.com

[2] Professor, Shri Vaishnav Institute of Forensic Science, Shri Vaishnav Vidyapeeth Vishawavidyalaya, Indore, Madhya Pradesh, India. E-mail: mpgoutam@gamil.com

Spectral Imaging technique which combines digital imaging with % reflectance spectroscopy for the analysis of writing ink, including gel pen inks. This technique facilitates the detection of reasonable differences where chemically resembling inks have to be examined, Authors have reported the analysis of a variety of blue, red and black gel inks on white office paper using this technique. They have compared analytical methods which are traditionally used for ink identification highlighting the discrimination potential for the red, blue and black gel inks. This method is claimed to be a non-destructive technique. In a study, forensic analysis of 33 inks from blue and black ballpoint pens of different models were examined and the discriminating power of UV-Vis, FTIR spectroscopy and TLC techniques were determined (Causin *et al.*, 2008). In view of recommendations and usefulness of this technique at different levels, the present study was under taken. Different types of documents were altered in the laboratory by same colour inks belonging to different companies and were examined under Visual Spectral Comparator for ink differentiations under filtered light.

MATERIALS AND METHOD

Documents of various natures were collected from different sources and these were altered by same color ink but belonging to different companies. Decipherment of the altered writing text was examined under VHC-6000HS (Foster & Freeman Co Ltd England). Analysis included Visible, Near Infrared multispectral imaging and Middle-Infrared hyper spectral imaging. Spectra measurement was carried out by selecting 400–1000 nm range. 10 simulated samples of Questioned documents were examined in which 16 different writing instruments were used.

Table 1: Selection of Writing Materials Used in the Ink Characterization

Group I Ballpoint Pen			*Group II Gel Pen*		
Sl. No.	*Color*	*Trade Name*	*Sl. No.*	*Color*	*Trade Name*
1.	Blue	Reynolds	9.	Blue	Montex
2.	Blue	Cello	10.	Blue	Rorito flymax
3.	Black	Flair	11.	Black	FlairPop
4.	Black	Luxor	12.	Black	Reynolds
5.	Red	Flair	13.	Red	Reynolds
6.	Red	Cello	14.	Red	Techno
7.	Green	Flair Matrix	15.	Green	Flair Matrix
8.	Green	Reynolds	16.	Green	Reynolds

OBSERVATIONS AND RESULTS

A. Original sample
B. Analyzed sample under VSC 6000
C. Spectral analysis of Ink

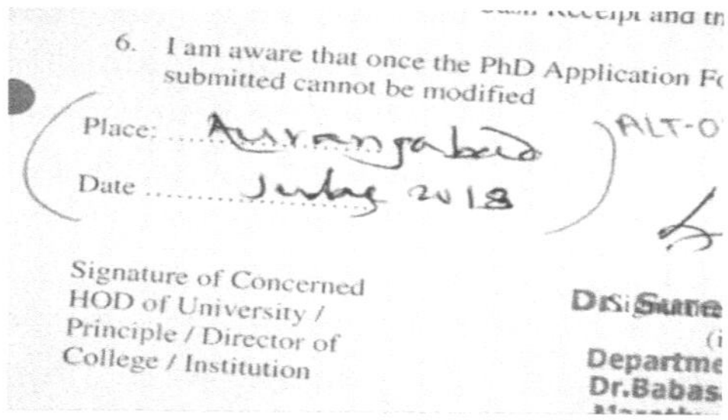

(a)

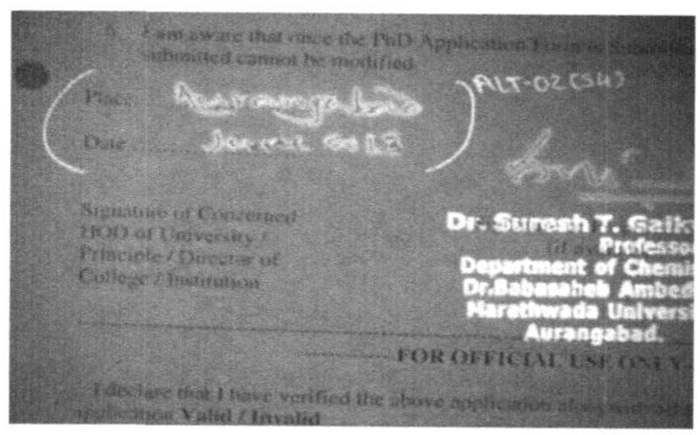

(b)

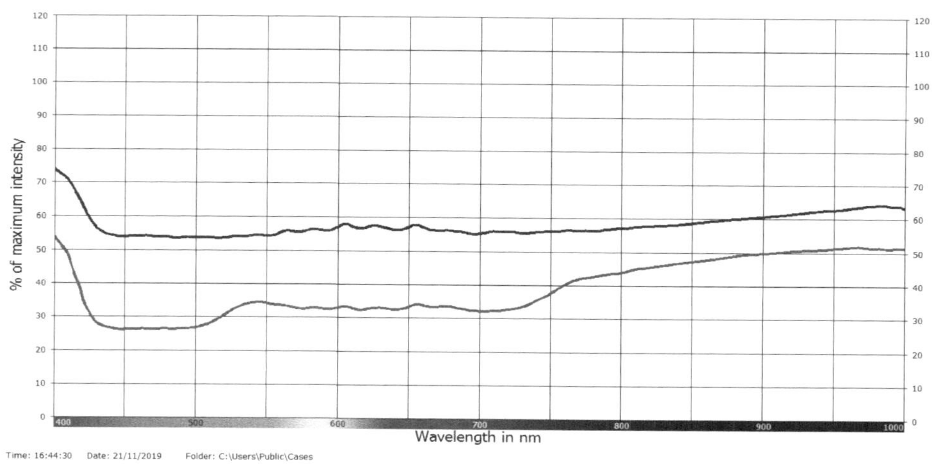

(c)

Fig. 1: Showed that the date and place was altered with two different ink, after examination, both ink showed different intensity of absorption vs wavelength. The λ_{max} value for Ink A and B is 540 nm and 610 nm, respectively. It indicates chemical composition of both ink are different as shown in Figure 1(c).

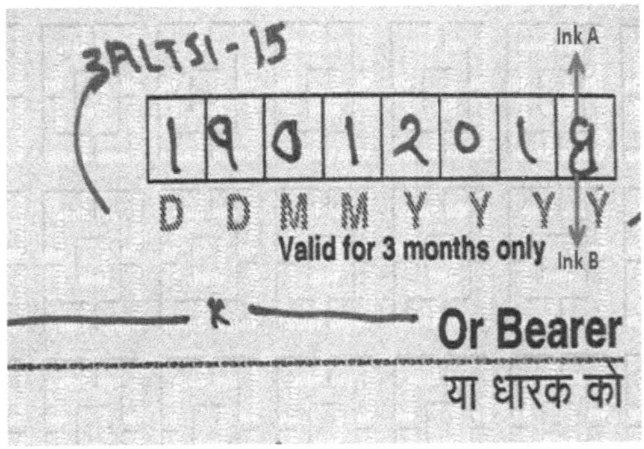

(a)

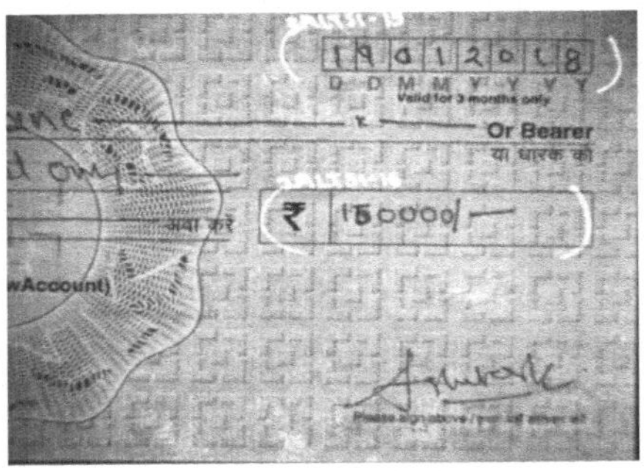

(b)

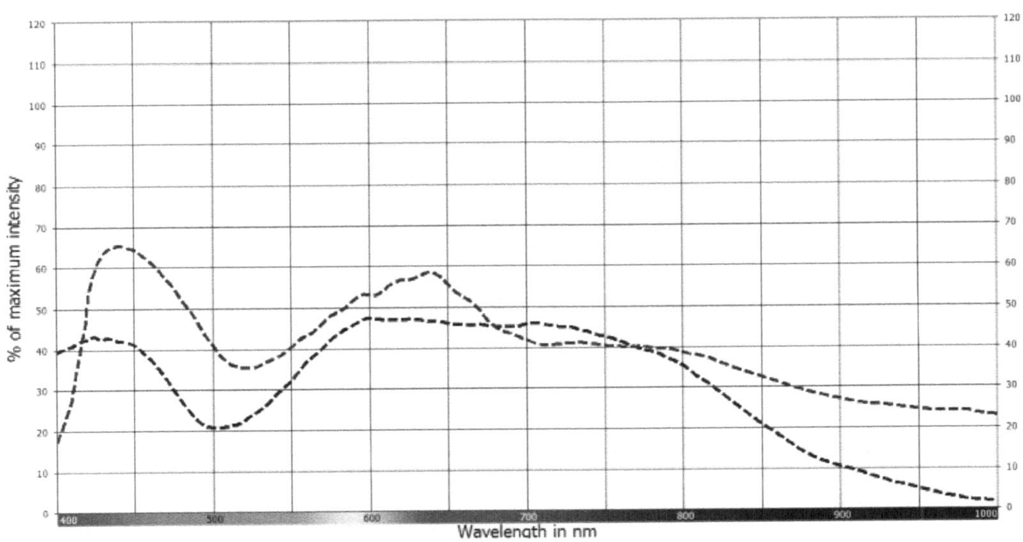

(c)

Fig. 2: Showed that the date on cheque was altered with different ink, after examination both ink showed different intensity of absorption vs wavelength. The λ_{max} value for Ink A and B is 420 nm and 600 nm respectively as shown in Figure 2(c).

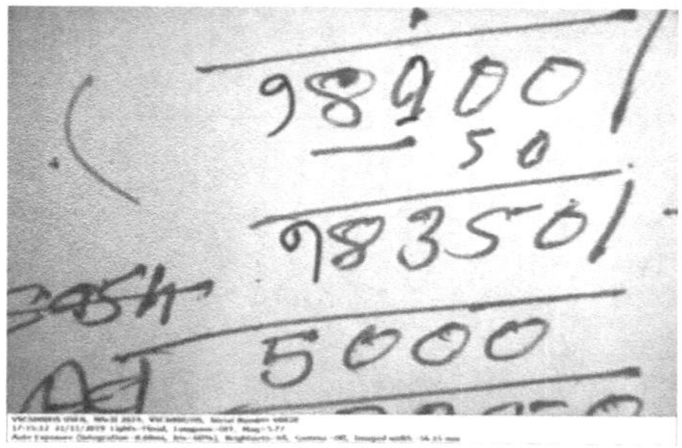

(a)

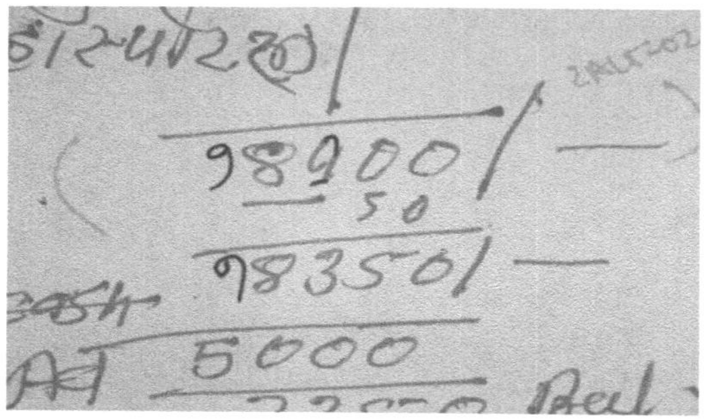

(b)

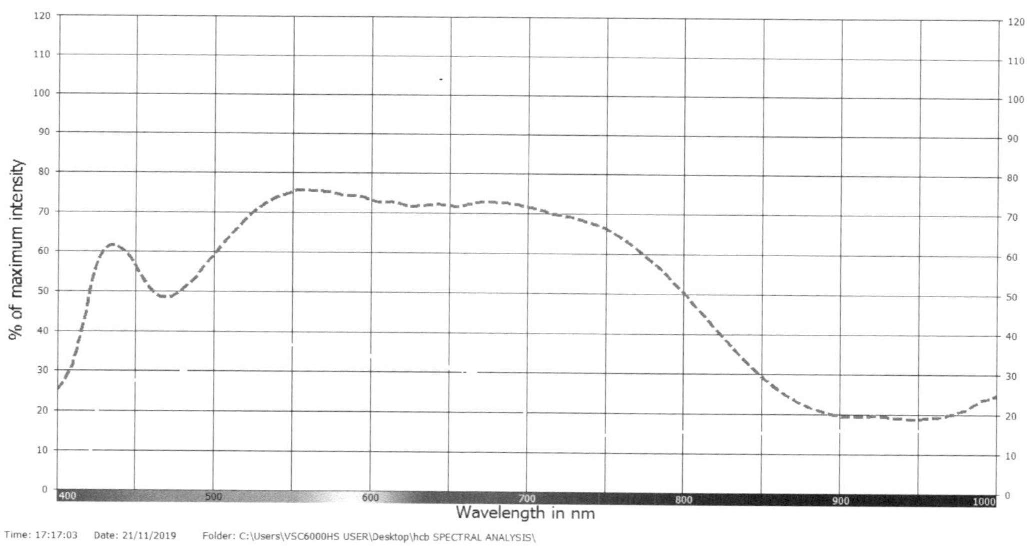

(c)

Fig. 3: Showed that the digit on purchase bill was altered with different ink, after examination Ink A and Ink B showed different intensity of absorption vs wavelength. The λ_{max} value for Ink A and B is 560 nm and 550 nm, respectively. Fig. 3(c) shows spectral differences of both the inks.

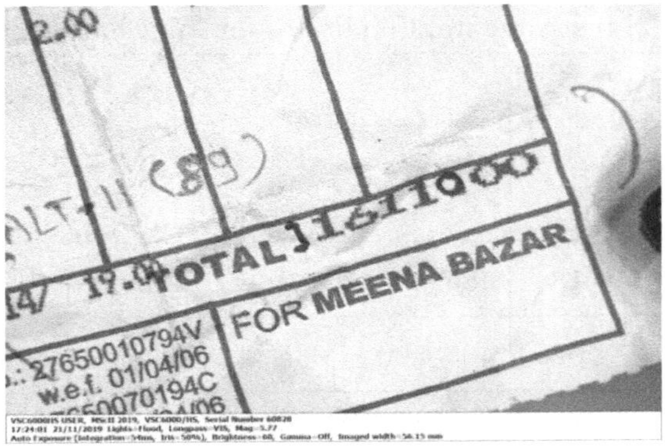

(a)

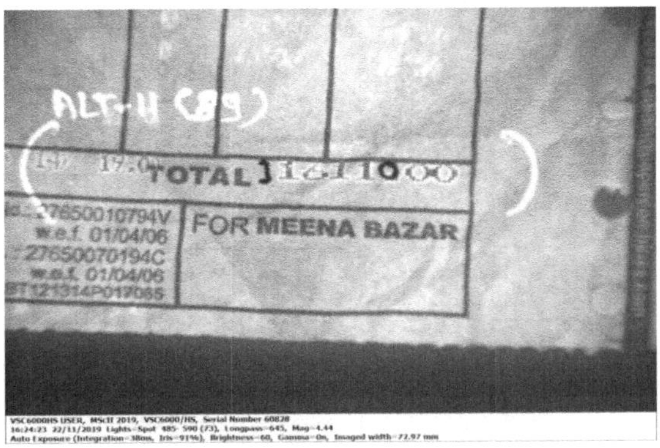

(b)

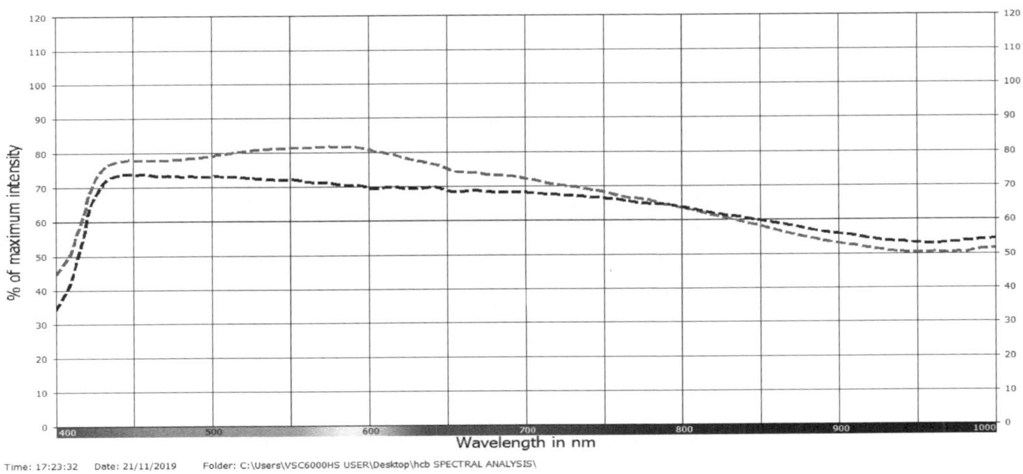

(c)

Fig. 4: Showed that digit on bill amount was altered with different ink, after examination both showed different intensity of absorption vs wavelength. The λ_{max} value for ink A and B is 590 nm and 440 nm respectively. Fig. 4(c) shows spectral differences of both the inks.

Table 2: Instrumental Conditions for Ink Characterization

Sample No.	*Writing Material*			*Light*	*Long Pass Filter*	*Auto Exposure*		*Gamma*	*Image Width*
	Color	*Nature*	*Sr. No.*			*Integration*	*Iris*		
1.	Red	Gel	13	485–590	645	24 ms	91%	on	56.74 mm
2.	Blue	Ball	02	485–590	645	79 ms	90%	–	75.88 mm
3.	Blue	Gel	10	485–590	695	83 ms	90%	on	75.88 mm
4.	Black	Ball	04	485–590	645	59 ms	90%	on	75.88 mm
5.	Black	Gel	11	flood	1000	1 s	50%	–	57.88 mm
6.	Black	Ball	03	485–590	645	21 ms	91%	on	43.78 mm
7.	Black	Ball	03	Spot	645	37 ms	90%	on	105.88 mm
8.	Black	Ball	03	485–590	645	43 ms	90%	on	115 mm
9.	Blue	Ball	01	485–590	645	25 ms	91%	on	64.54 mm
10.	Blue	Ball	01	485–590	645	38 ms	90%	on	64.54 mm

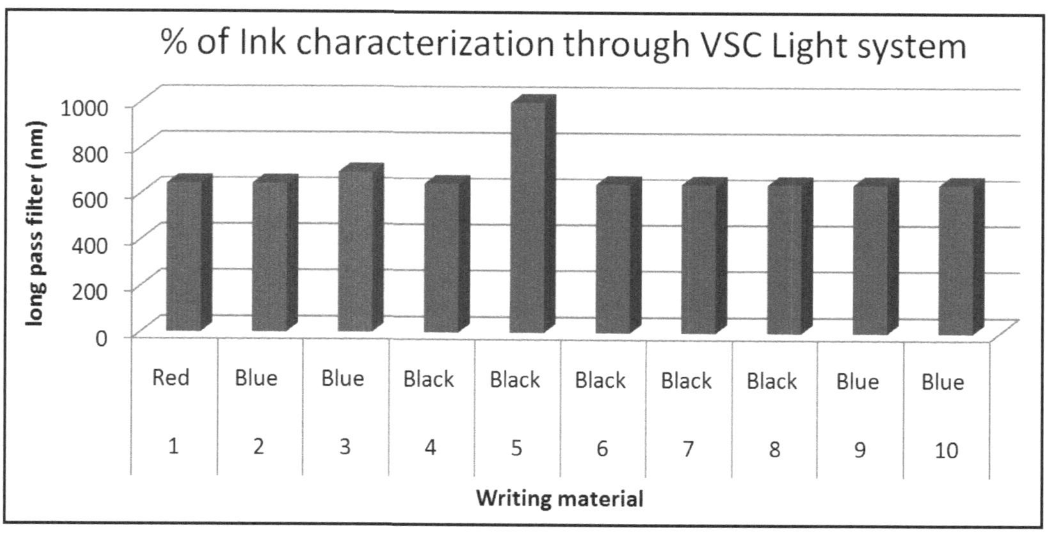

Graph 1: Showing the % of Ink Characterization through VSC (Table 2)

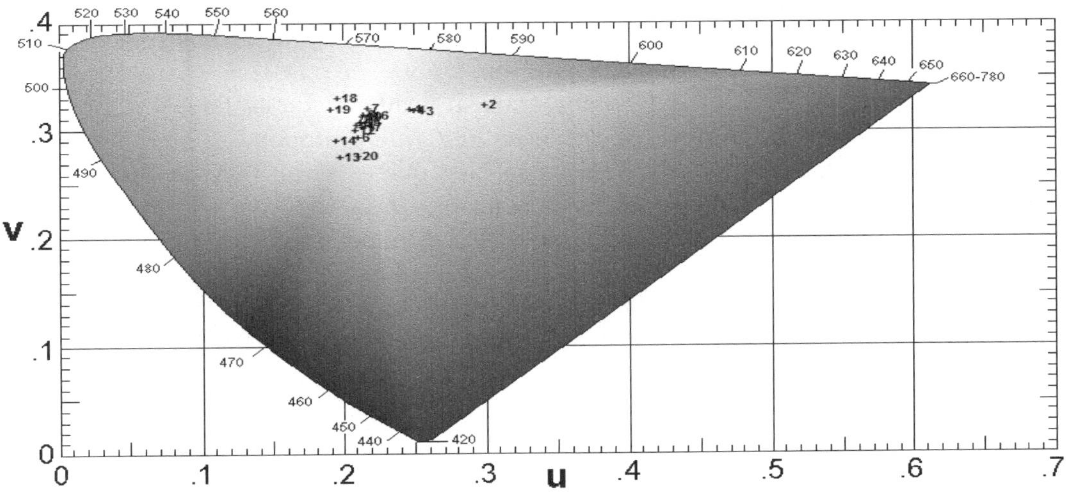

Graph 2: Showing the Spectral Analysis of the Ink Sample Carried on Spectral Range 490 m to 590 nm

CONCLUSION

Present study can provide an alternative tool for the characterization of ink. Better characterization could be possible at 485–590 nm. However, it is of interest to further study how factors such as ageing may affect the results. In the case of reference measurement, it is recommended that a blank reference should be used in order to minimize spectral interferences.

ACKNOWLEDGEMENT

The authors are thankful to the Director, Government Institute of Forensic Science, Ravindranath Tagore Marg, Civil Lines, Nagpur for the permission to conduct the research in Forensic Science Department and for his constant encouragement.

REFERENCES

[1] American Society for Testing and Materials: E444-07 (2007). *Standard Guide for Scope of Work of Forensic Document Examiners*, ASTM International, West Conshohocken.

[2] Causin, V., Casamassima, R., Marega, C., Maida, P., Schiavone, S., Marigo, A. and Villari, A. (2008). The discrimination potential of ultraviolet-visible spectrophotometry, thin layer chromatography, and Fourier transform infrared spectroscopy for the forensic analysis of black and blue ballpoint inks. *J Forensic Sci.*, 53*(6)*, 1468–73.

[3] Koppenhaver, K.M. (2007). *Equipping a laboratory, Forensic Document Examination: Principles and Practice*. New Jersey: Humana Press, 69–74.

[4] Lewis, J.A. (2014). *Instrumentation, Forensic Document Examination: Fundamentals and Current Trends*. Oxford: Academic Press, 69–80.

[5] Neumann, C. (2009). *Ink comparison and interpretation. A. Jamieson (Ed.),* Wiley Encyclopedia of Forensic Science, New Jersey. John Wiley & Sons.

[6] Reed, G., Savage, K., Edwards, D. and Nic Daeid, N. (2014). Hyperspectral imaging of gel pen inks: An emerging tool in document analysis. *Science & Justice 54*(1), 71–80.

[7] Siegel, J.A. (2013). Ink analysis in M.M. Houck, J.A. Siegel, P.J. Saukko (Eds.), *Encyclopaedia of Forensic Sciences*. Waltham: Academic Press, 375–379.

Constituent Factors of Brand Loyalty with Special Reference to Online Shopping

Surbhi Jajoo[1] **and Santosh Dhar**[2]

ABSTRACT

The Online market is flooded with several, leading national and global brands besides numerous small and local brands. Online shopping, despite divergent brands, is not well differentiated by the consumers. This results in fragmented market and leads to a highly competitive market. In Online market, strong brand loyalty and a wide distribution network are vital in attracting customers. Brand loyalty is built over a period of time by technological innovations, consistent high quality, aggressive advertisement and marketing. There remains a probability of brand switching due to impulsive buying. A deep insight into brand loyalty and satisfaction will facilitate marketers to retain the prevailing customers and stimulate new ones. In this backdrop, this study is undertaken to examine constituent factors of brand loyalty with respect to online shopping.

Keywords: Aggressive Advertisement, Connect, Fondness, Impulsive Buying, Trust.

INTRODUCTION

In today's, business scenario e-marketing is the fastest growing segment of online trade. The major distinction between retail and online selling is the intensity of communication between the consumer and the seller. There is much electronic interactivity with the customer through emails and FAQs. Through FAQs, the consumer's queries on payment, policies, shipment, product and other matters can be addressed effectively. It is especially important for businesses to understand the customer loyalty because it affects long term growth of the organizations. It can be said that to determine the success of the businesses, the customer satisfaction and loyalty are the important parameters.

"Brand loyalty has been one of the biggest issues in the world of marketing" (Kotler, 1994). Additional attention was put on creating loyalty when marketers

[1] Research Scholar, Shri Vaishnav School of Management, Shri Vaishnav Vidyapeeth Vishwavidyalaya, Indore, Madhya Pradesh, India. E-mail: surbhi.jajoo@gmail.com

[2] Dean, Faculty of Doctoral Studies and Research, Shri Vaishnav Vidyapeeth Vishwavidyalaya, Indore, Madhya Pradesh, India. E-mail: deanresearch@svvv.edu.in

recognized that brand loyalty has a strong relation to increased sales, gives companies the option of premium pricing, and encourages the customers to search for their preferred brand (Giddens & Hoffman, 2002). "Additional factor leading to improved brand loyalty is that it is six times costlier to earn a new buyer compared to retaining an existing one" (Rosenberg & Czepiel, 1984). Loyal customers not only spread positive word of mouth about their preferred brand in fact, they not only defend the brand in arguments and discussions, but also talk down competitive brands. "This implies that brand loyalty not only influences the brand positively but can have negative effects on the competitive brands" (Raju *et al.*, 2009).

The growth of an organization depends on its skill to entice consumers towards brands. It is important to retain existing customers and make them loyal to the brand for growth of the organization. It is said that trading with loyal customers is less costly than attracting new customers. The brand loyal consumer does not evaluate product attribute but simply chooses the familiar brand based on positive feelings towards it. The overall positive evaluation stems from past experience, happiness with the brand and satisfaction enjoyed with the particular brand under consideration. The brand loyalty is generated usually when, a customer makes a trial purchase of a specific product from a given brand. If the product meets the customer's expectations and gives satisfaction, the consumer often forms a habit of buying this same product or service. The customer is now interested in not only buying the product again and again but other products too from the same company because the brand is now considered safe and satisfactory.

Customers who are loyal stick to a brand, and are ready to pay an extra price for a specific brand over others and readily recommend the given brand to other people (Giddens & Hofmann, 2002). Keller (2008) has defined brand switching as brand jumping, referring to the processing which a customer change from buying one brand of a product to buying another brand. According to Van Trip *et al.* (1996), majority brand switches occur not because they are inherently rewarding but because they are essential for attaining or avoiding another purchase or consumption. Young adults are observed to be easily attracted by such motivation because of inadequate financial resources (Ness *et al.*, 2002). Furthermore, Shukla (2009) established a relationship between brand switching and brand loyalty and found that brand switching occurs due to decline in brand loyalty and growing acceptance of other brands, which leads to customer's willingness to try alternative brands.

In general, brand loyalty can be described as fondness for a brand compared to other same options. It is usually evaluated through repeat purchase. Brand loyalty can be defined as a consumers' choice, stated through intention or behaviour to repurchase a brand on everyday basis. "With competition increasing day by day, customer maintenance and growth has become the first goal of many companies" (Chegini, 2010) and "loyal customers can be considered as a key to success" (Tripathi, 2009). There is a need to identify various kinds of factors that leads to loyalty.

REVIEW OF LITERATURE

Jacoby and Chestnut (1978) contended that brand loyalty is a function of psychological (decision-making, evaluative) processes. Brands are chosen according to internal criteria resulting in a commitment towards the brand, which is an essential element of brand loyalty. This definition identifies six requirements for brand loyalty biased, social response, expressed over time, by some decision-making unit, about one or more alternate brands out of a group of such brands, and a function of emotional (decision-making, evaluative) processes.

Harris and Goode (2004) tested a loyalty model including four determinants: service quality, satisfaction, perceived value, and trust in online services (book and ticket purchasing). Their study revealed 16 valid and reliable measurement items of brand loyalty dimensions for both online and offline services. Rawly and Dawes (1999) stated that Brand Loyalty is the likelihood of positive attitudes and behaviours of consumers towards a particular brand and this could amount to repeat purchase and positive word-of-mouth. Brand loyalty has been stated to be the supreme goal of marketing.

Lau *et al.* (2006) identified seven factors responsible for shift in consumers' brand loyalty. The factors were: promotion, style, store atmosphere, brand name, product quality, price, and service quality. Popular brand names lead to product benefits and higher recall of advertised benefits than non-popular ones. Janson-Boyd (2010) explained that need may be established through the culture in which we are raised. On giving incentives on repeat purchase of the product or service consumer may become brand loyal.

Dick and Basu (2014) stated that customer loyalty generates positive word of mouth communication and aggressive strategies are avoided by loyal customers. Such results persuade strategists and marketers to develop and maintain strong customer loyalty. Kotler and Keller (2005) stated that "on the basis of 20–80 principle, 80% of profitability of a company is created by 20 % of satisfied customers". Thus, a favourable connection between a company and its customers is lucrative for the business.

Bennet (2011) explained that some drivers of brand loyalty are inertia, involvement, perceived risk, satisfaction, habit and relationship between product or service providers. A number of positive coincidences will increase trust, relationship commitment customer satisfaction and continuity. Ozer (2005) perceived that corporate image, service quality, trust and customer switching costs are the influential factors of brand loyalty. Also, the results of the study illustrated that trust is one of the crucial antecedents of brand loyalty. However, perceived switching cost and perceived service quality seemed to have the same level of impact on brand loyalty (Shih *et al.*, 2004). Sufficient numbers of researches have advocated the role of factors like acceptance of technology and online service quality playing a significant role to influence customer satisfaction and loyalty towards online shopping.

Shrivastava and Dixit (2017) found that customers with longer association on online platforms with any service provider have greater loyalty. Longer association can only be possible if customers are satisfied with overall service offerings. There are research studies which show negaitive correlation between brand loyalty and brand switching. Sharma (2017) stated that the SERVQUAL dimensions are used to measure the quality of service provided by the website designs and the factors associated with the website usage to evaluate customer e-satisfaction which further leads to customer e-loyalty.

OBJECTIVES

- To develop and standardize tool to measure the brand loyalty among Online Customers.
- To identify constituent factors of Brand Loyalty of Online Customers.

METHOD

Study: It is an exploratory research and sample universe is existing online customers.

Design: A multi-stage randomised design was used for this study.

- *Stage 1*: An operational definition was framed based on review of literature.
- *Stage 2*: Eighty-five (85) statements were made keeping in view the operational definition.
- *Stage 3*: The statements were reviewed and repetitive as well as ambiguous statements were dropped. 42 items were selected for getting validated from judges. 20 judges were executives working in Branding and marketing field with minimum experience of 5–7 years and 10 academics with work experience of 10–15 years.
- *Stage 4*: The statements that received more than 75 percent of acceptance by the judges were retained for developing the scale.
- *Stage 5*: Seventeen (17) statements were selected.
- *Stage 6*: A tool of 17 statements, measuring on a five-point Likert scale was finalized.

Sample: The tool developed in this study was administered on 450 online customers. After collecting the data, 18 filled up forms were dropped being incomplete. The data was collected using simple random sampling from Indore, Mumbai, Ahmedabad and Pune city of India.

Tools for Data Collection: The tool developed for the study was administered for data collection. The tool comprises of 17 statements measured on a five-point Likert scale.

Tools for Data Analysis: Standardization is a mandatory process in developing a tool, hence reliability and validity norms were computed for standardization of this scale. SPSS 23.0 version was used for data analysis. Item-total correlation and Factor analysis were employed for analysing the data. Spearman's half split reliability test was computed for the scale and it was found to be 0.71. Validity of the test is checked which indicated high validity on account of being 0.84.

RESULTS AND DISCUSSION

Item-total correlation: Item total correlation was computed to determine the internal consistency among the items. All the items were found significant with value more than 0.1946. Thus, the scale was finalised on 17 items.

Factor Analysis: KMO-Bartlett test for sample adequacy was computed on the data and the score came out to be 0.656 which is acceptable. The data was then subjected to factor analysis and five factors: Trust, Connect, Convenience, Fondness and Expectations.

Factor 1: Trust

The factor Trust is constituted of four items with a total factor load of 3.22 and 21.51% of variance. The items are as under:

Trust	I am loyal towards online shopping site I use.	.914
	My preference for shopping site will not change.	.824
	I prefer the shopping site I use compared to other.	.818
	To switch from my current shopping site will need major rethinking.	.673

Trust is defined "as the level of reliability guaranteed by one party to another within a given exchange relationship". Trust is developed when you offer relevant service according to the customer's need. Customer trust in a company can play a considerable role in defining the customer's actions towards company. "Customer trust (a belief) influences customer intentions (Ajzen, 1991)". "Empirical research has shown that trust increases customer intention to purchase a product from a company (Jarvenpaa *et al.*, 2000) as well as customer intention to return to a company (Doney and Cannon, 1997)".

Factor 2: Connect

The factor Connect is constituted of four items with a total factor load of 2.75 and 16.95% of variance. The items are as under:

Connect	I am happy with most of the online shopping sites that I use, so I am unlikely to look for alternatives.	.786
	Commercials regarding competing brands does not change my mind to switch the shopping site I use.	.706
	I cannot change my usual shopping site even if another site is offering heavy discount.	.672
	I am willing to pay a price premium to purchase from same shopping site again.	.593

Connect means "to associate or relate in some respect". Connected customers turn out to be repeat customers, and repeat customers are important for business success. Involvement and attitudinal loyalty are highly correlated (Park 2006). A consumer typically decides to undergo a repeat purchase when he/she is normally satisfied with the brand and is in the last stage of post purchase behavior of the consumer buying process (Kotler & Armstrong, 2016). Connected customers may not need to be persuaded to come back, they come back because they want to. They make regular and frequent purchases and become supporters for the brand.

Factor 3: Convenience

The factor Convenience is constituted of three items with a total factor load of 2.33 and 12.62% of variance. The items are as under:

Convenience	In case my present online shopping site is not available, I can wait for it.	.918
	I will continue using same online shopping sites that I have been using.	.894
	I will stick to my shopping site even if a better site comes up.	.520

Convenience means "to proceed without difficulty and add to one's ease or comfort". If customers are stymied and frustrated in their efforts to seek information or consummate transactions, they are less likely to come back (Cameron, 1999). Convenience in service can be thought of as a means of adding value to consumers by decreasing the amount of time and effort a consumer must expend on the service (Colwell *et al.*, 2008). Convenience to a customer is an element of customer experience that saves the customer's time and effort. It can also be understood as the ease of operating any product or services without frustration.

Factor 4: Fondness

The factor Fondness is constituted of three items with a total factor load of 1.85 and 8.99% of variance. The items are as under:

Fondness	I feel attached to particular shopping site over others.	.776
	I say positive things about the shopping site I use	.755
	I recommend this shopping site to my friends and relatives.	.327

Fondness means "affection or liking for something". Fondness of a customer is the liking of the product or shopping site which make them to praise the site. Customers are so inclined to the site that they are unable to switch from the same. "Brand commitment is the link between the customer self and a brand, where the customer thinks that the preferred brand is an integral part of his life" (Walsh *et al.*, 2010). "Brand commitment is the emotional, psychological, and economic connection that a consumer may have towards a brand, with the expectation to create a long-term relationship" (Chang and Wu, 2014). "Positive intentions to maintain and strengthen a business relationship are created by commitment, thus customer loyalty is positively influenced by commitment" (Cater, 2010). Customers who are fond of a particular product are asset to the company as they disseminate positive things which in return increases the customers.

Factor 5: Expectation

The factor Expectation is constituted of three items with a total factor load of 2.10 and 7.66% of variance. The items are as under:

Expectation	I have been using same online shopping site for many years now.	.768
	Popularity of shopping site matter to me.	.677
	Shopping site I use meets my expectations.	.660

Customer expectation "is experience about a product, service or a brand that a customer's hold in their mind". Customer expectations are any set of behaviours or actions that individuals anticipate while interacting with a company. According to Pizam (2010) customer expectations are the beliefs about service delivery that serve as standards for the judgment of the actual service delivered to the customer. Customer expectations are designed in the minds of customers on the basis of their experiences, combined with their pre-existing experience and knowledge.

CONCLUSION

The study is undertaken to develop a tool to measure Brand loyalty and identify the constituent factors that lead to Brand loyalty with special reference to Online shopping. The tool is comprised of 17 items on a five-point Likert scale, varying from 'strongly disagree' to 'strongly agree'. The split half reliability of test is 0.71 and the validity is 0.84. Five factors namely Trust, Connect, Convenience, Fondness and Expectation were identified and discussed with the available literature.

IMPLICATIONS

The tool developed in this study can be used by researchers and marketers to assess the brand loyalty. Marketers can use the results of this study to develop their marketing strategy. Factors identified in this study can be used to assess the exact reason behind the increased brand loyalty. This study can be treated as a preliminary study to further conduct an exhaustive research on a larger and more diverse set of population.

REFERENCES

[1] Ajzen, I. (1991). The theory of planned behavior. *Organizational Behavior and Human Decision Processes*, 50(2), 179–211.

[2] Aydin, Serkan, Özer and Gökhan (2005). The analysis of antecedents of customer loyalty in the Turkish mobile telecommunication market. *European Journal of Marketing*, 39, 910.

[3] Cameron, Michelle (1999). Content that works on the web, *Target Marketing*. 1 (November), 22–58.

[4] Cater, T. and Cater, B. (2010). Product and relationship quality influence on customer commitment and loyalty in B2B manufacturing relationships. *Industrial Marketing Management*, 39(8), 1321–1333.

[5] Chang, H.H. and Wu, L.H. (2014). An examination of negative e-WOM adoption: Brand commitment as a moderator. *Decision Support Systems*, 59, 206–218.

[6] Chegini, M.G. (2010). Customer loyalty and Value key dimensions interaction in Organization. *China-USA Business Review*, 9(8), 8–14.

[7] Colwell, S.R., Aung, M., Kanetkar, V. and Holden, A.L. (2008). Toward a measure of service convenience: multiple-item scale development and empirical test. *Journal of Services Marketing*, 22 (2), 160–169.

[8] Dick, A.S. and Basu, K. (1994). Customer loyalty towards an integrated framework. *Journal of the Academy of Marketing Science*, 22(2), 99–113.

[9] Giddens, N. and Hofmann (2010). *Brand Loyalty*. Iowa State University. Working paper.

[10] Jacob, Jacoby and David, B. Kyner (1973). Brand Loyalty vs. Repeat Purchasing Behavior. *Journal of Marketing Research*, 10, 1–9.

[11] Jansson-Boyd, C.V. (2010). Consumer psychology. Berkshire, England: McGraw-Hill Education.

[12] Jarvenpaa, S.L., Tractinsky, N. and Vitale, M. (2000). *Information Technology And Management*, 1(1/2), 45–71.

[13] Keller and Kevin, L. (2008). *Strategic Brand Management*. Englewood Cliffs, NJ: Prentice Hall.

[14] Kiran, Sharma and Saptarshi, Das (2017). Service Quality and Customer Satisfaction – With Special focus on the Online Cab Industry. *India International Journal of Business and Management*, 12(7):192.

[15] Kotler, P. (1984). Marketing Management: Analysis, Planning and Control. Englewood Cliffs: Prentice-Hall.

[16] Kotler, P. and Armstrong, G. (2016). *Principles of Marketing*. Global Edition, 16th edn, Pearson.

[17] Kotler, P. and Keller, K.L. (2005). *Marketing Management* 12th edition New Delhi Prentice-Hall of India.

[18] Lau, Mm; Chang, Mt; Moon, Kl and Liu, Ws (2006). The brand loyalty of sportswear in Hong Kong. *Journal of Textile and Apparel, Technology and Management*, 5(1).

[19] Lloyd, C. Harris∗, Mark, M.H. Goode (2004). The four levels of loyalty and the pivotal role of trust: a study of online service dynamics. *Journal of Retailing*, Vol. 80, pp. 139–158.

[20] Ness, M., Gorton, M. and Kuznesof, S. (2002). The student food shopper: segmentation on the basis of attitudes to store features and shopping behaviour. *British Food Journal*, 104(7), 506–525.

[21] Patricia, M. Doney and Joseph, P. Cannon (1997). An Examination of the Nature of Trust in Buyer-Seller Relationships. *Journal of Marketing*, 61(2), 35–51.

[22] Pizam, A. (2010). Creating memorable experiences. *International Journal of Hospitality Management*, 29(3), 343.

[23] Raju, Sekar, Unnava, H. Rao, Montgomery, Nicole Votolato (2009). The Moderating Effect of Brand Commitment on the Evaluation of Competitive Brands. *Journal of Advertising* 38 (2), 21–35.

[24] Rosenberg, L.J. and Czepiel (1984). A Marketing Approach for Customer Retention. *Journal of Consumer Marketing* 1(2), 45–51.

[25] Rowley, J. and Dawes, J. (1999). Customer loyalty – a relevant concept for libraries. *Library Management*, 20(6), 345–351.

[26] Rundle-Thiele, S. and Bennett, S. (2001). A brand for all seasons: A discussion of brand loyalty approaches and their applicability for different markets. *Journal of Product & Brand Management*, 10(1), 25–37.

[27] Shih, H.P. (2004). An empirical study on predicting user acceptance of e-shopping on the web. *Science Direct*, 351–368.

[28] Shukla, P. (2009). Impact of contextual factors, brand loyalty and brand switching on purchase decisions. *Journal of Consumer Marketing,* 26(5), 348–357.

[29] Srivastava, Shaila and Dixit, Shweta (2017). "Brand Switching – A Case of Mobile Telecom Industry in India. Economics and Applied Informatics, "Dunarea de Jos" University of Galati, *Faculty of Economics and Business Administration*, 1, 48–53.

[30] Tripathi, M.N. (2009). Customer Engagement-Key to Successful Brand Building. Vilakshan: *The XIMB Journal of Management,* 6, 131–140.

[31] Van, Trijp, H.C.M., Hoyer, D.W. and Inman, J. (1996). Why switch? Product category-level explanations for true variety-seeking behavior. *Journal of Marketing Research*, 33(3), 281–292.

[32] Walsh, M.F., Winterich, K.P. and Mittal, V. (2010). Do logo redesigns help or hurt your brand? The role of brand commitment. *Journal of Product & Brand Management*, 19(2), 76–84.

Deep Learning Algorithm for Cloud-Based Architecture for Smart Agriculture

Avdesh Kumar Sharma[1] **and Vivek Dubey**[2]

ABSTRACT

Agriculture is unique in its processes for quality food production. Unfortunately, agriculture's efficiency is precisely low, and numerous factors affect the produce. Cloud-Based Architecture (CBA) with deep learning is one of the solutions for the above problem. CBA procedures site gives the crop management perception based on dignified data using sensors and data analytics to discover the root reason for yield reduction. Precision agriculture automates farming, including gathering data and investigating for improved decision-making and to increase the yield and superiority of the cloud-based agricultural product. The CBA combined with data analytics and deep learning is so-called smart farming or smart agriculture. The objective of intelligent agriculture is to grow a decision-making provision system for farming management. The precision smart agriculture can be improved through the benefit of IoT's latest technologies and proposed Novel Deep Learning Algorithm.

Keywords: Cloud-Based Architecture, Deep Learning, Internet of Technology, Smart Agriculture.

INTRODUCTION

Smart agriculture is an emerging indicator of the perceptions since IoT sensors provide evidence in agriculture fields. Agriculture these days has been automated by utilizing automatic programming robotics. The present paper attempts to highlight that the agrcloud can entirely be contingent on propagating seed and can be observed by Arm processor and Cloud based IoT agriculture. It is mainly planned to limit the effort of farmers. The development in various technologies in this contemporary era has ensured that the reduction of sensors, and the efforts to exploit them in numerous ranges are accomplished. Efforts have been put to use IoT technology in an agricultural area to develop innovative

---

[1]Research Scholar, Shri Vaishnav Institute of Information Technology, Shri Vaishnav Vidyapeeth Vishwavidyalaya, Indore, Madhya Pradesh, India. E-mail: avdeshsharma@svvv.edu.in

[2]Assistant Professor, Shri Vaishnav Institute of Information Technology, Shri Vaishnav Vidyapeeth Vishwavidyalaya, Indore, Madhya Pradesh, India. E-mail: vivek.dubey@svvv.edu.in

farming resolutions (Rezk, *et al.*, 2021). IoT has transported an excessive revolution in the agriculture situation by investigative numerous difficulties and tasks in farming (Hafeez *et al.*, 2019). Through the improvement of technology, it has been predicted that efforts of IoT agriculturalists and technologists can discover the solutions to the problems such as shortage of water, cost management, and efficiency issues, which farmers face (Permana *et al.*, 2019). State-of the-art IoT technologies can address these issues and provide solutions for intensification, efficiency, while dropping the cost.

Determinations completed on wireless sensor networks permit us to gather data commencing sensing devices and send it to the central servers (Naman *et al.*, 2019). Data collected done sensors gives information around different environmental conditions to monitor the whole system appropriately. Observing the environmental circumstances or crop productivity is the factor for estimating crop, but frequent other factors affect the crops' efficiency. According to Moazzam *et al.* (2020) field supervision, soil and crop monitoring, attacks of uninhabited animals, thefts, etc. are other factors that affect the crop efficiency. Agricultural network architecture, stage, and topology, can benefit IoT support and enable farmers to improve crop efficiency. The present paper summarizes current and continuing improvements in IoT agricultural applications, devices/sensors, communication protocols, and numerous pioneering technologies.

Numerous critical dimensions of IoT-based agriculture, counting technologies, industry trends, and countries' policies have been adressed to satisfy innumerable stakeholders. It is equally vital that many big organizations have started investing in developing novel methods for farm management systems expanding IoT. Finally, the standard data classification expending CNN (Convolution neural network) model has been used in finding and prediction. IoT data classification technology has been used in current years to contribute through agriculture data classification to regulate agricultural construction and intensification of crop production. The use of the smart device in artificial intelligence and to provide a situation for pest identification, completing a combination of environmental factor evidence and deep learning to support farmers to examine crop growth trends.

REVIEW OF LITERATURE

Ponnusamy *et al.* (2021) determined and utilized advanced technologies for smart farming. Mohapatra *et al.* (2021) proposed a well-organized IoE based smart agricultural model (IoESA) as an innovative AgriTech tool for the proficient message between all IoT-based devices. It also presented a sensor-based soil quality monitoring technique called Soil-Smart Agriculture (SSA) to benefit the growers' growth in modern soil quality. Saha *et al.* (2021) examined the machine learning and meta-heuristics methods giving different illustrative application domains of IoT and thereby developing a learning pattern with an added advantage. Rezk *et al.* (2021) proposed an intelligent technique based on the mixture of a wrapper feature collection method, and PART classification method for crop efficiency and drought prediction. The consequences specified that the projected process is intense, precise to classify and calculate crop efficiency and deficit compared to the prevailing techniques.

Hafeez *et al.* (2019) predicted that accuracy is better through the use of MMI method and FCRBM model. On the different side, the convergence rate is improved by GWDO algorithm. Simulation significances prove that the proposed fast and precise model outperforms existing models. Namani *et al.* (2019) presented a Smart Drone for crop management wherever the real-time Drone data joined through IoT and Cloud Computing technologies building a maintainable Smart Agriculture. Used the VGG19 process for image feature extraction and recognition in the discovery of 24 types of pests from crop images. It was detected and different pests were classified using several CNN approaches and concluded that the classification significances complete machine learning methods such as SVM and Fisher.

DEEP LEARNING TO IDENTIFY AND CLASSIFY

Deep learning (DL) is based on the machine learning (ML) framework. The training procedure primarily includes unsupervised learning (UL) and clusters the training data set to acquire what kind of data will be categorized. Supervised learning (SL) is then achieved to label respectively entry's predictable output value and complete the feature vectors in the training data set definite as input.

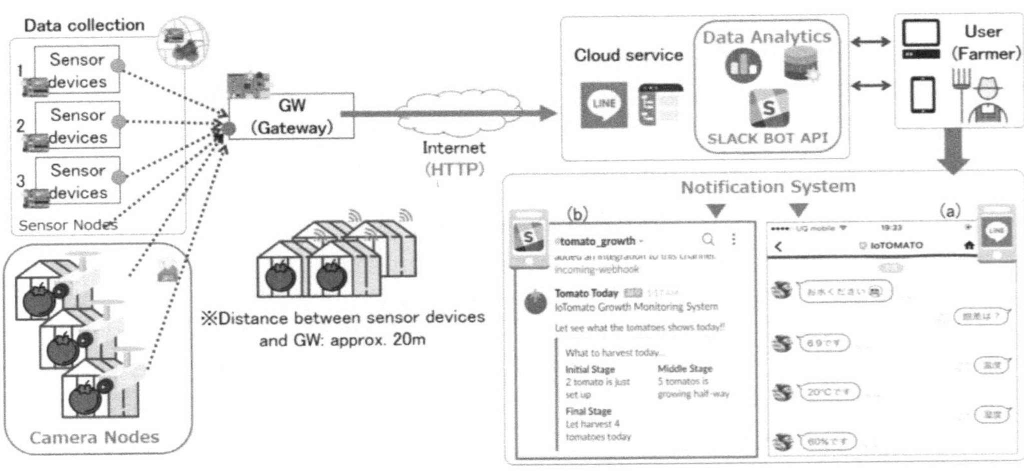

Fig. 1: Deep Learning for Cloud-Based Architecture (CBA) Data Analysis
(*Source:* Ponnusamy & Natarajan, 2021)

The forecast classification is defined as the output. Lastly, the loss function estimates the standard deviation between the predictable consequence and the concrete production. There are two collective methods in deep learning for crop growth data identification and classification:

1. The VGG19 technique for image dataset extraction features extraction and recognition in detecting 24 types of pests.
2. Perceived and confidential dissimilar pests exploiting numerous CNN means and connected the classification signified through machine learning (ML) methods such as SVM and Fisher. The machine learning method's classification accuracy was 79%, and the classification correctness rate of the CNN and RNN technique was 90%.

DEEP LEARNING FOR CLOUD-BASED ARCHITECTURE (CBA) DATA ANALYSIS

The deep learning model (DLM) is presently the maximum usually used model in RNNs (Recurrent Neural Networks) which is collected of four major components: the output gate, the forget gate, input gate, and memory cell.

1. *Input Gate:* After a feature is assumed as an input, the input gate controls whether or not to input the worth in this repetition.
2. *Memory Cell:* Stores the intended value to be second hand in the subsequent stage.
3. *Output Gate:* Controls whether or not to crop the predictable value in this recurrence.
4. *Forget Gate:* Controls whether or not to robust the memory, analogous to a recommence. The mathematical building of the RNNs model. The input is signified as g(z). The input gate is f (zi); the over-all activation function f usages the sigmoid function to control the probability of turning on the gate. The likelihood that the input gate is twisted on can be determined by cumulative g(z) and f(zi). The memory cell archives the present value of the input, enhances the preceding input value, and increases the probability of the forget gate to choose whether or not to overlook the before recorded value [c' = g(z)f (zi) + cf (zf)]. The value of ct−1 is a vector stored in the RNNs model memory, anywhere the vector input at time t is zt. A linear conversion function increases zt to find g(zt) as an input. zt is multiplied by another linear conversion function to get I(zt) to control the input gate. zt is multiplied by another linear conversion function to obtain F(zt) to control the forget gate, and zt is multiplied by a linear conversion to get O(zt) to control the output gate (The RNNs model). The value of g(zt) is reproduced by the value of I(zt) after the Sigmoid function. The consequence is added to F(zt) product subsequently the Sigmoid function and ct−1 and stored in the RNNs model. The value of O(zt) after the Sigmoid function is increased by the

stored value to get the output value zt. Similar steps are surveyed to compute the worth to be stored in the RNNs model and the output value after the subsequent input data zt+.

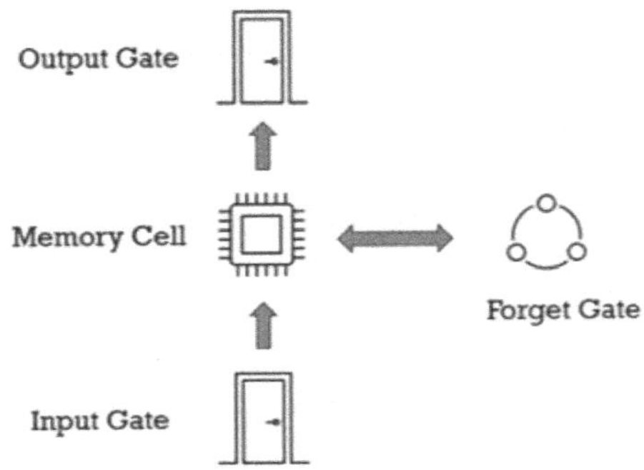

Fig. 2: Deep Learning for Cloud-Based Architecture (CBA) Data Analysis

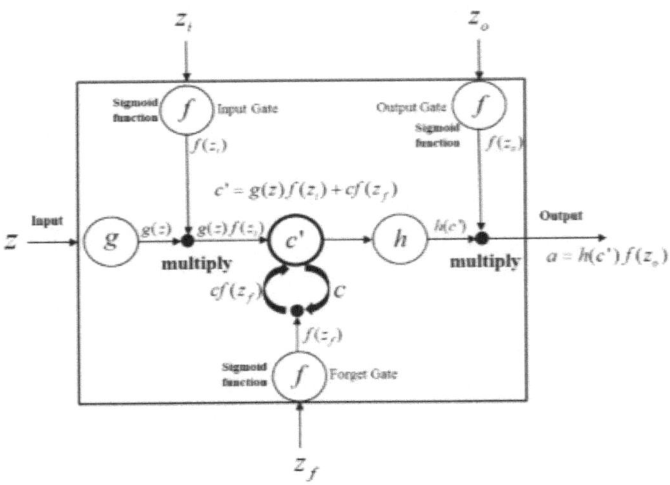

Fig. 3: Working of RNNs Model

CLOUD INFRASTRUCTURE

Cloud provides a huge quantity of storage done in large virtualized servers associated for achieving necessary action. A Cloud-based IoT strategy has been accessible in for precision farming (Rezk *et al.*, 2021). IoT techniques are functional to investigate and accomplish data from farms complete sensors and devices to produce decision-making information. Enterprises recommend IoT agricultural network based on Cloud. Cloud storage layer centralize completely agriculture-related data such as weather-related, soil, impregnation, crop, and agricultural promotion in the Cloud and offers on-demand resources and finished networke infrastructure. Analytics possessions and web services are similarly connected on Cloud or the internet nearby by cloud services. Maximum of the devices or sensors are not intended to attach through the internet for data sharing. To manage data sharing problems, indigenous gateways are intended to attach hardware devices and sensors for connectivity, security, and controllability altogether.

Implementation of the gateway in the conservatory or arena advances automation and controls the real-time conservatory monitoring system (CMS). Hardware modules and cloud services are dispersed, whereas resources are integrated with complete fog computing. Fog computing diminishes the computational load of the Cloud and guarantees real-time dispensation. The rudimentary resolve of fog computing in this proposed network stage influences cloud computing reserves' on-demand scalability by the attractive benefit of composed Cloud and edge computing. In hardware modules, multiple actuators, sensors, microcontrollers also central processing units have been practical to monitor and sense frequent agricultural variables. Hardware fundamentals are distributed in wide or local networks to produce services or actions.

SENSOR DISTRIBUTION

Environmental sensors complete a wireless environmental sensor transmission stage and an extensive cloud data computing server system to gather environmental data. These data were examined to regulate the real-time agricultural meteorology, soil conditions, crop growth, and additional applicable information essential to assist the environment's supervision. The research and growth of wireless sensor transmission systems emphasizes on manifold sensors to communicate services, complete a wireless network in actual and send the information posterior to the system for examination. It is one of the indispensable growth applications of the IoT in smart agricultural technology (SAT). In this study, the sensor components were consistently arranged in farmland on a hillside, rendering the wireless transmission sensor appearances. The environmental data composed by the sensor modules were combined and transmitted to the cloud database completed a wireless communication protocol for subsequent analysis and calculation. Consistently distributed sets of environmental sensing components in the farmland, to detect in dissimilar environments. the intended environmental sensing component were obtained.

ANALYZING ENVIRONMENTAL SENSOR DATA WITH RNN

A total of five sets of conservational sensor information were collected. Data for respectively conservational sensor components was composed at an interval of one-hour. Respectively, set of data comprised real-time conservational details for the time, temperature, humidity, light, soil humidity, appropriate pressure, and comparative altitude. Till date, 10000 sets of environmental data have been collected. The environmental data had insignificant difference due to the place of the sensor component. Further, RNN was used to investigate the six sensor components to predict conservational factors at five locations and used deep learning approaches to predict whether environmental factors have consequence in presence. The proposed approach combined an agricultural supervision mobile application to deliver pest image collection, real-time pest identification, and sensor data visualization. The developed mobile application delivers real-time access to environmental data and environmental sensors' historical records. The environmental data's analytical consequences can be displayed displayed in an App's procedure on the mobile device and delivers farmers with real-time agricultural management.

CONCLUSION AND FUTURE WORK

Multiple sensors are placed to collect environmental data and to deliver immediate information on damage in numerous farmland areas so that the assembly amongst occurrence and connected environmental factors can be resolved. RNN neural network was used to construct an apposite training model constructed on an on-site study of the sternness in by professional farm managers and the environmental data. The model is efficient to examine the predicted data and regulate the extent of pest incidences. The anticipation and control information is then provided to the farmers to reduce the agricultural damage caused by the known pests as speedily as probable. A bigger size of dataset can be used to improve and find higher accuracy in the detection system. Besides, the classification model can be further enhanced. Using a CNN and RNN and uniting analysis with daily data sensing values such as temperature, humidity, illuminance, etc will help to detect. Lastly, detected regions of images can be used to further detect and classify the diseases appearing either on leaf or on sections. The present work can deliver this IoT explanation for an automatic system to provide greenhouse farmers with better crop management strategies.

REFERENCES

[1] Alonso, R.S., Sittón-Candanedo, I., Casado-Vara, R., Prieto, J. and Corchado, J.M. (2020). Deep Reinforcement Learning for the management of Software-Defined Networks in Smart Farming. In 2020 *International Conference on Omni-layer Intelligent Systems (COINS),* 1–6.

[2] Bhanu, K.N., Mahadevaswamy, H.S. and Jasmine, H.J. (2020). IoT based Smart System for Enhanced Irrigation in Agriculture. In 2020 *International Conference on Electronics and Sustainable Communication Systems (ICESC),* 760–765.

[3] Boukhris, L., Abderrazak, J.B. and Besbes, H. (2020). Tailored Deep Learning based Architecture for Smart Agriculture. In 2020 *International Wireless Communications and Mobile Computing (IWCMC),* 964–969.

[4] Chatterjee, A. and Basu, S. (2019). Green Sense: A Smart Assistant for Agriculture Management using IoT and Deep Learning. In 2019 *6th International Conference on Computing for Sustainable Global Development (INDIACom)* (pp. 495–499). IEEE.

[5] Davcev, D., Mitreski, K., Trajkovic, S., Nikolovski, V. and Koteli, N. (2018). IoT agriculture system based on LoRaWAN. In 2018 *14th IEEE International Workshop on Factory Communication Systems (WFCS),* 1–4.

[6] Hafeez, G., Javaid, N., Riaz, M., Ali, A., Umar, K. and Iqbal, Z. (2019). Day ahead electric load forecasting by an intelligent hybrid model based on deep learning for smart grid. In *Conference on Complex, Intelligent, and Software Intensive Systems,* 36–49.

[7] Hati, A.J. and Singh, R.R. (2020). Towards Smart Agriculture: A Deep Learning based Phenotyping Scheme for Leaf Counting. In *2020 International Conference on Smart Technologies in Computing, Electrical and Electronics (ICSTCEE),* 510–514.

[8] Mekala, M.S. and Viswanathan, P. (2017). A novel technology for smart agriculture based on IoT with cloud computing. In 2017 *International Conference on I-SMAC (IoT in Social, Mobile, Analytics and Cloud)(I-SMAC),* 75–82.

[9] Moazzam, S.I., Khan, U. S., Tiwana, M.I., Iqbal, J., Qureshi, W.S. and Shah, S.I. (2019). A Review of Application of Deep Learning for Weeds and Crops Classification in Agriculture. In *2019 International Conference on Robotics and Automation in Industry (ICRAI),* 1–6.

[10] Mohapatra, H. and Rath, A.K. (2021). IoE based framework for smart agriculture. *Journal of Ambient Intelligence and Humanized Computing,* 1–18.

[11] Murakami, Y. (2014). iFarm: development of web-based system of cultivation and cost management for agriculture. In 2014 *Eighth International Conference on Complex, Intelligent and Software Intensive Systems,* 624–627.

[12] Murakami, Y., Utomo, S.K.T., Hosono, K., Umezawa, T. and Osawa, N. (2013). IFarm: development of cloud-based system of cultivation management for precision agriculture. In *2013 IEEE 2nd global conference on consumer electronics (GCCE),* 233–234.

[13] Namani, S. and Gonen, B. (2020). Smart Agriculture Based on IoT and Cloud Computing. In 2020 *3rd International Conference on Information and Computer Technologies (ICICT),* 553–556.

[14] Permana, I.K.Y.T., Mantoro, T., Irawan, E., Handayani, D.O.D. and Safitri, C. (2019). Increased efficiency of Smart Water Systems for Vegetable Plants Using the Deep Learning Classification Approach. In 2019 *5th International Conference on Computing Engineering and Design (ICCED),* 1–5.

[15] Ponnusamy, V. and Natarajan, S. (2021). Precision Agriculture Using Advanced Technology of IoT, Unmanned Aerial Vehicle, Augmented Reality, and Machine Learning. In *Smart Sensors for Industrial Internet of Things,* 207–229.

[16] Rajeswari, S., Suthendran, K. and Rajakumar, K. (2017). A smart agricultural model by integrating IoT, mobile and cloud-based big data analytics. In 2017 *international conference on intelligent computing and control (I2C2),* 1–5.

[17] Rezk, N.G., Hemdan, E.E.D., Attia, A.F., El-Sayed, A. and El-Rashidy, M.A. (2021). An efficient IoT based smart farming system using machine learning algorithms. Multimedia Tools and Applications, 80(1), 773–797.

[18] Ribeiro, F.M., Prati, R., Bianchi, R. and Kamienski, C. (2020). A Nearest Neighbors based Data Filter for Fog Computing in IoT Smart Agriculture. In 2020 *IEEE International Workshop on Metrology for Agriculture and Forestry (MetroAgriFor),* 63–67.

[19] Saha, A., Chowdhury, C., Jana, M. and Biswas, S. (2021). IoT Sensor Data Analysis and Fusion Applying Machine Learning and Meta-Heuristic Approaches. Enabling AI Applications in Data Science, 441–469.

[20] Suciu, G., Istrate, C.I. and Dițu, M.C. (2019). Secure smart agriculture monitoring technique through isolation. In 2019 *Global IoT Summit (GIoTS),* 1–5.

[21] Tombe, R. (2020). Computer Vision for Smart Farming and Sustainable Agriculture. In 2020 *IST-Africa Conference (IST-Africa),* 1–8.

[22] Wang, Y. (2020). Research on Construction and Management of Intelligent Agriculture Cloud Platform. In 2020 *International Conference on Computer Information and Big Data Applications (CIBDA),* 340–343.

[23] Xu, X.K., Li, X.M. and Zhang, R.H. (2019). Remote Configurable Image Acquisition Lifting Robot for Smart Agriculture. In 2019 *IEEE 4th Advanced Information Technology, Electronic and Automation Control Conference (IAEAC),* 1, 1545–1548.

[24] Yan, H., Yu, P. and Long, D. (2019). Study on deep unsupervised learning optimization algorithm based on cloud computing. In 2019 *International Conference on intelligent transportation, Big data & smart city (ICITBS),* 679–681.

[25] Zhu, J., Yu, T., Zheng, S., Niu, C., Gao, J. and Tang, J. (2020). Hemp Disease Detection and Classification Using Machine Learning. In 2020 *International Conferences on Internet of Things (iThings) and IEEE Green Computing and Communications (GreenCom) and IEEE Cyber, Physical and Social Computing (CPSCom) and IEEE Smart Data (SmartData) and IEEE Congress on Cybermatics (Cybermatics),* 878–887.

Determination of Gender using Different Features of Handwriting

Hansi Bansal[1] **and Swati Dubey Mishra**[2]

ABSTRACT

Handwriting is considered to be the written speech of an individual which may differentiate him/her from other individuals. It is proved that handwriting is brain writing which is controlled by the signals sent to the nervous system through our hands. Handwriting has been considered as an important tool for the identification of the suspect. Different attributes of handwriting have been useful in determining the gender of the individual. Handwriting is identified by the combination of all elements and characteristics related with all forms of letters obtained from the hand movement. In this study, 100 handwriting samples have been studied to differentiate gender on the basis of different features of handwriting of the individuals. Handwriting samples were obtained from 50 males and 50 females aged between 18–30 years having a basic knowledge of writing English. Informed consents were obtained from the subjects. Different features of handwriting both micro and macro were examined. The data obtained was subjected to statistical analysis for determining the significant results.

Keywords: Gender, Handwriting, Suspect, Writing Movement.

INTRODUCTION

Examination of handwritten documents for determination of the identity of the writer has a great importance in the provision of justice. Over the years, many cases have been dealt with evidence provided by different handwriting documents (Ahuja *et al.*, 2018). Handwriting is an important evidence which is analysed for the authentication of a questioned document. It is an acquired skill that is a complex neuro muscular task formed by the co-ordinated movement under the neural network of movements of arm, hand and fingers. With practice and skill over the time, the formation of handwriting becomes automatic and less subjected to control of consciousness (Huber, 1999, Kelly & Lindblom, 2006, Kumar *et al.*, 2013).

[1] Research Scholar, Shri Vaishnav Institute of Forensic Science, Shri Vaishnav Vidyapeeth Vishwavidyalaya, Indore, Madhya Pradesh, India. E-mail: hansi.fsc@gmail.com

[2] Assistant Professor, Shri Vaishnav Institute of Forensic Science, Shri Vaishnav Vidyapeeth Vishwavidyalaya, Indore, Madhya Pradesh, India. E-mail: swati.dubeymishra@gmail.com

Handwriting pertains to two fundamental fields that include i) the study of handwriting as a neuromuscular activity, its development as a skill & different factors affecting it. ii) the study of handwriting identification as a discriminatory process (Huber, 1999, Kumar *et al.*, 2013). Discrimination in handwriting is obtained by the comparison of writing habits and a thorough examination of similarities and differences in various elements of handwriting. These discriminatory elements are commonly referred as features or characteristics of handwriting. The writing characteristics which are simply the manifestations of the habits formed are classified as class and individual characteristics (Morris, 2020; Nunez, 2020; Upadhyay *et al.*, 2017). Class characteristics are developed as a result of prescribed writing systems whereas individual characteristics are the particular idiosyncrasies of an individual (Sahu & Ya, 2017 and Hilton, 1992). Gender Identification has developed as an important area and gained much attention due to its significant applications in varied fields especially in forensics (Topaloglu & Ekmekci, 2017; Bi *et al.*, 2019). Gender determination from handwriting is important for the examiners to narrow the suspect's pool (Bi *et al.*, 2019; Desai & Kalyan, 2013).

This paper aims to analyse the handwriting characteristics for the determination of gender from samples collected from the population of Central India.

METHOD

The study was conducted on 100 handwriting samples (50 males and 50 females) obtained from population from central India. Subjects were chosen randomly for the handwriting samples aged between 18–40 years having a basic knowledge of reading and writing English. Written informed consents were taken from the subjects. All the demographic and personal details like name, education, etc. were noted down. Samples were taken on an A-4 size sheet having the contents of the Standard London letter. Subjects were asked to sit comfortably with stable seating arrangement and copy the content on the given sheet in their normal speed of writing with sufficient amount of light. People suffering with any deformity of hand or in any unhealthy condition that could affect the formation of handwriting, were excluded from the study. Samples were scanned, micro and macro features of handwriting were observed for gender identification.

RESULTS AND DISCUSSION

Data was obtained from hundred samples of males and females. Handwriting features were extracted and examined for studying the gender variations. The analysis depicted the rejection of null hypothesis; there is no significant difference between the handwritings of males and females. There is a positive significant differences in the writings of males and females. Figure 1 depicts the format of the sample collected from the subjects. Table 1 shows the frequency of occurrence (in percentage) of males and females. Significant differences were obtained after the statistical analysis was done. Figure 2 represents the examined handwriting characteristics. Table 2 depicts the p values obtained from Z test (two-tailed). Values marked with * represents the significant values at 0.05 significant level. Handwriting characteristics marked with * were found to represent significant gender variations.

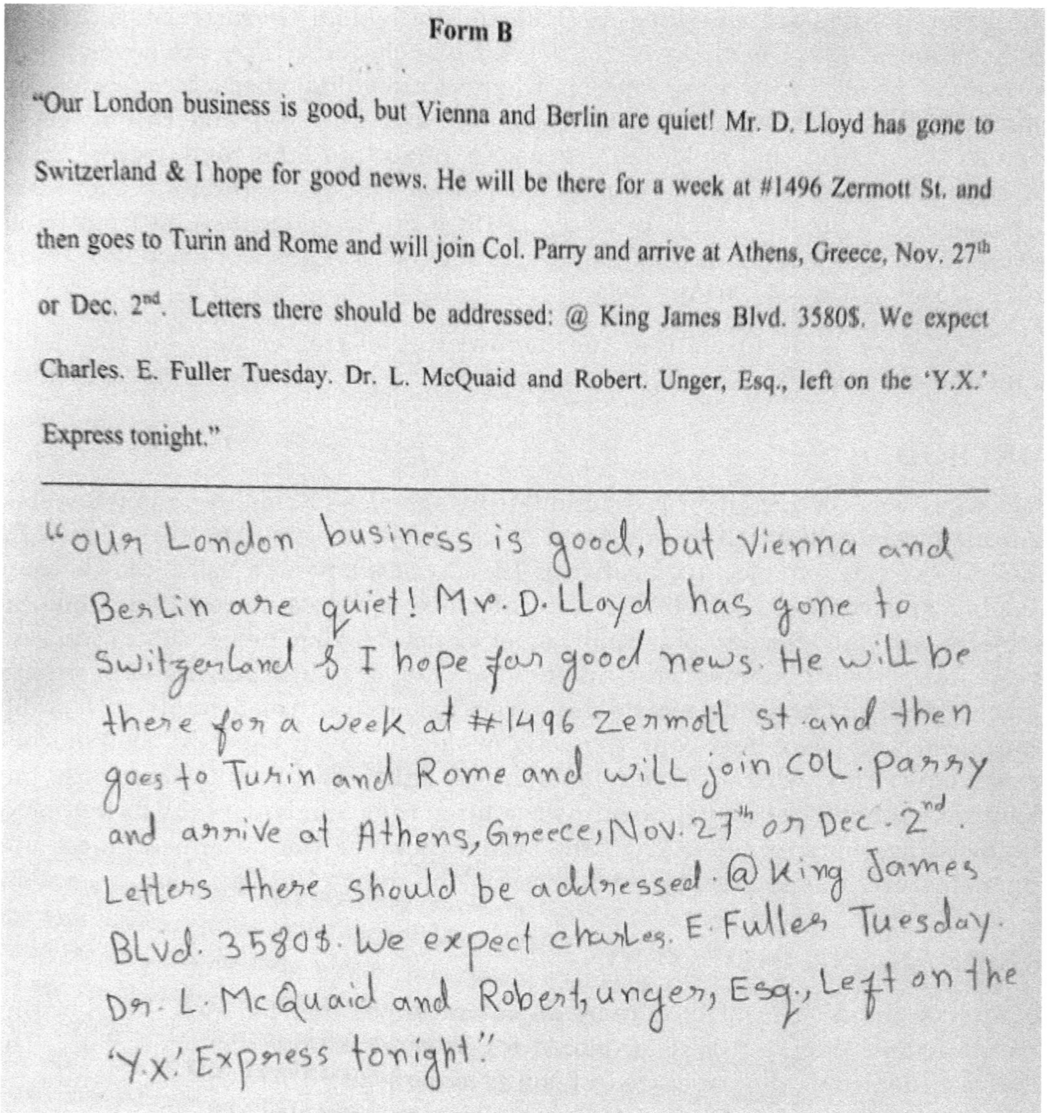

Form B

"Our London business is good, but Vienna and Berlin are quiet! Mr. D. Lloyd has gone to Switzerland & I hope for good news. He will be there for a week at #1496 Zermott St. and then goes to Turin and Rome and will join Col. Parry and arrive at Athens, Greece, Nov. 27th or Dec. 2nd. Letters there should be addressed: @ King James Blvd. 3580$. We expect Charles. E. Fuller Tuesday. Dr. L. McQuaid and Robert. Unger, Esq., left on the 'Y.X.' Express tonight."

"our London business is good, but Vienna and Berlin are quiet! Mr. D. Lloyd has gone to Switzerland & I hope for good news. He will be there for a week at #1496 Zermott st. and then goes to Turin and Rome and will join COL. parry and arrive at Athens, Greece, Nov. 27th or Dec. 2nd. Letters there should be addressed. @ King James BLvd. 3580$. We expect charles. E. Fuller Tuesday. Dr. L. McQuaid and Robert, unger, Esq., Left on the 'Y.X.' Express tonight."

Fig. 1: Handwriting Sample

Table 1: Frequency of Occurrence of Handwriting Characteristics

S. No.	*Characteristics of Handwriting*	*Frequency of Occurrence in Males (%)*	*Frequency of Occurrence Inn Females*
1.	Cursive Writing	4	12
2.	Dispersive Writing	5	11
3.	Dot over 'i'	82	68
4.	Stem of Letter 'b'	8	6
5.	Consistency in 't' crossbars	75	62
6.	Eyelets	7	11
7.	Formation of 'r'	51	56
8.	Upwards stroke of 's'	50	55
9.	Base of 'w' (Angular)	90	88
10.	Initial stroke of 'a'	30	32
11.	Direction (Anticlock-wise)	2	3
12.	Retracing	6	8
13.	Buckles	21	15

Table 2: *p*-Values Obtained after a Two-Tailed Z-test

S. No.	*Examined Characteristics of Handwriting*	*p-Value*
1.	Cursive Writing	0.029*
2.	Dispersive Writing	0.101
3.	Dot over 'i'	<0.00001*
4.	Stem of Letter 'b'	0.56
5.	Consistency in 't' crossbars	<0.00001*
6.	Eyelets	0.298
7.	Formation of 'r'	<0.00001*
8.	Upwards stroke of 's'	<0.0001*
9.	Base of 'w' (Angular)	<0.00001*
10.	Initial stroke of 'a'	0.681
11.	Direction (Anticlock-wise)	0.645
12.	Retracing	0.562
13.	Buckles	0.21

*Significant at 0.05 level.

Fig. 2: Examined Characteristics of Handwriting

CONCLUSION

Identification of handwriting is important evidence in the field of forensic science. Examination of handwriting for gender identification has evolved as a great research area in the field of forensic science and has proved its significance. This study has shown the significant differences in the writing of males and females by analysing micro and macro features of handwriting which will be helpful to identify the gender at preliminary level.

REFERENCES

[1] Ahuja, P., Chaudhary, P.K., Manve, P., Christal, G. and Dahiya, M.S. (2018). Study of Genetic and Environmental Influence on Handwriting of Monozygotic Twins. *Indian Journal of Scientific Research*, 17–22.

[2] Bi, N., Suen, C.Y., Nobile, N. and Tan, J. (2019). A multi-feature selection approach for gender identification of handwriting based on kernel mutual information. *Pattern Recognition Letters, 121*, 123–132.

[3] Desai, B. and Kalyan, J.L. (2013). Forensic Examination of Handwriting and Signatures. *International Journal of Innovative Research and Development 2*(5), ISSN 2278–0211.

[4] Hilton, O. (1992). *Scientific examination of questioned documents*. CRC press.

[5] Huber, R.A. and Headrick, A.M. (1999). *Handwriting identification: facts and fundamentals*. CRC press.

[6] Kelly Seaman, J. and Lindblom, B. (2006). Scientific examination of questioned documents. *New York, USA*, 133. ISBN: 10: 0-8493-2044-5

[7] Kumar, S., Saran, V., Vaid, B.A. and Gupta, A.K. (2013). Handwriting and gender: A statistical study. *Z Zagadnień Nauk Sądowych*, (95), 620–626.

[8] Morris, R.N. (2020). *Forensic handwriting identification: fundamental concepts and principles*. Academic press. ISBN: 0-12-507640-1

[9] Núñez, F., Maraver, M.J. and Colzato, (2020). L.S. Sex Hormones as Cognitive Enhancers? *J. Cogn Enhanc* 4, 228–233. https://doi.org/10.1007/s41465-019-00156-1

[10] Sahu, M. and Ya, A. (2017). Examination of Handwriting. *Journal of Current Research, 9*(01), 45923–45928.

[11] Topaloglu, M. and Ekmekci, S. (2017). Gender detection and identifying one's handwriting with handwriting analysis. *Expert Systems with Applications, 79*, 236–243.

[12] Upadhyay, S., Singh, J. and Shukla, S.K. (2017). Determination of Sex Through Handwriting Characteristics. *Int J. Cur. Res. Rev.,* Vol, 9(13), 11.

Effect of Angular Writing Surface Station on Alignment of Handwriting

A Forensic Study

Anurag Sahu[1] **and M.P. Goutam**[2]

ABSTRACT

Handwriting of a person is an acquired skill, which is controlled by brain and muscles; it is a neuro-muscular phenomena. Writer learns to copy a book form of a particular language according to his/her own appropriateness and implements it in own handwriting. Since childhood, a person learns handwriting skills on a horizontal surface station (flat) with forearm movement because it is easiest for muscles of hand. Once muscles adopt the neuromuscular pattern of handwriting, writing becomes more smooth, speedy and perfect. But in some conditions writer can get inclined surface stations (angular/inclined) for writing. In such situations handwriting may be different from that on the horizontal surfaces. Therefore, the handwriting of the same writer on varying inclined surface stations (10 degree to 90 degree) were examined against handwriting on horizontal surface stations. Handwriting samples were analysed under VSC-6000/HS for changes in alignment and it was observed that alignment of inclined surface station's handwriting was found up slanted, down slanted, mixed and zigzagged from straight alignment comparative to horizontal station's handwriting of the same writer. This study concludes that the forensic document examiners should not overlook the significant factor i.e., writing surface area/station of the document under study.

Keywords: Alignment Variations, Effect on Handwriting, Handwriting, Inclined Surface Station.

INTRODUCTION

The writing written by hand is known as handwriting. But in some exceptional cases handwriting cannot be produced by hand, such writer use their foot or mouth or both for writing. Therefore, handwriting is the reproduction of mental images which are stored in our memory and executed through neuron and muscular coordination. At the learning stage, handwriting is the conscious act but

[1]Research Scholar, Shri Vaishnav Institute of Forensic Science, Shri Vaishnav Vidyapeeth Vishwavidyalaya, Indore, Madhya Pradesh, India. E-mail: anuraggifs@gmail.com

[2]Professor, Shri Vaishnav Institute of Forensic Science, Shri Vaishnav Vidyapeeth Vishwavidyalaya, Indore, Madhya Pradesh, India. E-mail: mpgoutam@gmail.com

becomes subconscious after the repeated use of letters and words of a particular language. Smooth, speedy and superior line quality signs are found in handwriting of a graphically matured writer. Since childhood, the blueprint of neuromuscular pattern of handwriting in the brain makes no two people write exactly alike in an extended handwriting sample (Koppenhaver, 2007). Individual or unique handwriting characteristics in an individual's handwriting is distinguishable from other handwriting. In addition, natural variation is also an important part of an individual's handwriting.

Movement of writing is also an important aspect of handwriting and it could be a fundamental question in handwriting examination. Out of all the movements, forearm or muscular movement is best because writing through this is rapid, clear cut and smooth. Easiest, most rapid and most perfect writing is produced with the forearm movement used in connection with a slight action of the hand and fingers (Osborn, 1910). Since childhood, most people have learned handwriting skill on horizontal writing surface stations (flat surfaces) with forearm movement because it is easiest for muscles of the hand. Once muscles adopt the neuromuscular pattern of handwriting, writing becomes more smooth, speedy and perfect. Writers are accustomed with horizontal writing surface stations at most of the places such as school, college, home, office etc. But in some conditions a writer can get inclined writing surfaces (angular stations) or unusual surfaces for execution of writing and in such situations handwriting can be different from the horizontal writing surface stations.

Wall *et al.* (1991) carried out study to examine the effect a desk with a 10° inclination on their sitting posture during writing and reading. Subjects were analysed and compared for both flat and inclined desks. Marschall *et al.* (1995) conducted a study to compare muscular activity levels and sitting posture of young children while seated on a traditional workstation and ergonomically designed workstation (sloping desk). Shena *et al.* (2002) conducted a study to examine the effect of ergonomic desk design for improving motor accuracy in the writing performance of cerebral palsy students. Review of literature indicates that there is an impact on the handwriting characteristics of a writer due to the inclined workstation. Therefore, more research is needed with different aspects. The present study has focused on 09 inclined surface station's handwriting and observed the variations of alignment (a handwriting character) while comparing with the handwriting of horizontal writing surface stations of the same writer.

MATERIALS AND METHOD

Subject

A professional person of age of 32 years was selected for participation in this study. An informed consent was taken from the subject.

Experimental Setup

An angular slant board was placed on table and a subject was asked to take a seat on a non-moving chair by keeping thighs parallel to each other and hips well positioned. Hips and knees were at 90° with flat feet on the floor. Right elbow was resting on a slant board and the left elbow was resting upon the lap of the subject. White light source was maintained by switching on tube lights in the room for visibility.

Sample Preparation

The subject was asked to write a paragraph of around 75 words on A-4 white and unrolled paper having 70 GSM (B2B Copier) placed on an angular slant board. An original content was placed in front of the subject on the same table and a black ball pen of Cello Company, Fine grip model, bearing no. 03 19 HC 70015 H on refile was selected for handwriting. A total 10 handwriting samples from same subject were produced (09 samples of inclined surface and 01 of horizontal surface) on different angular surface station varying from 0° to 90°.

Angular Slant Board

Specifically designed slant board was used for writing at different angular surfaces. The surface of slant board tilted from 0° to 90° in anticlockwise direction. With the help of adjusting knobs, placed on the protractor shape structure on the left side, the slant board was fixed at desired angle (Figure 1a and 1b).

Sl. No.	*Angular Slant Board*	
	(a) Front View	*(b) Side View*
1.		

Fig. 1: (a) Front View and (b) Side View of Angular Slant Board having A-4 Size White Paper on Slant Board Adjusted at 20°

Video Spectral Comparator-6000/HS

PC based VSC-6000/HS instrument (Foster and Freeman Company Limited, UK.) with a wide monitor and VSC software was used in this study.

RESULTS AND DISCUSSION

Examination of handwriting samples of same writer under VSC-6000/HS, variations have been found in alignment while compared to horizontal (0°) and inclined surface (10° to 90°) station's handwriting (Table 1). As the writing surface station became angular, alignment changed in up-slanted, zigzagged, mixed and down slanted from straight. From the result it is concluded that the frequency of occurring up-slanted alignment is more in inclined surface station's handwriting when compared to mixed, zigzagged, down slanted and straight. Straight alignment remained constant and unchanged at 0° and 10°. Zigzagged alignment remained constant and unchanged at 50° and 60°. Up-slanted alignment was found at 20°, 40° and 90°. Downward alignment was observed only at 80°.

Table 1: Alignment Variations in Handwritings at Different Angles of Working Station

Sl. No.	*Handwriting at Various Angular Writing Surface Station*	*Variations in Alignment*
1.	0°	Straight
2.	10°	Straight
3.	20°	Up-slanted
4.	30°	Mixed
5.	40°	Up-slanted
6.	50°	ZigZagged
7.	60°	ZigZagged
8.	70°	Mixed
9.	80°	Down slanted
10.	90°	Up-slanted

CONCLUSION

Study concludes that the forensic document examiners should not overlook the significant factor i.e., writing surface area/station of the document under study for comparison of individual's writing.

ACKNOWLEDGEMENT

The authors are thankful to the Director and Head of Department of Forensic Science, Govt. Institute of Forensic Science, Aurangabad, M.H., for supporting, guiding and providing a platform to accomplish this project.

REFERENCES

[1] Koppenhaver, K.M. (2007). Forensic document examination: Principles and practice. Humana Press Inc. 7.

[2] Marschall, M., Harrington, A.C. and Steele, J.R. (1995). Effect of work station design on sitting posture in young children. *Ergonomics*, 38(9), 1932–1940.

[3] Osborn, A.S. (1910). Questioned documents, Rochester N.Y., The lawyers co-operative publishing co. 108.

[4] Shen, I., Kang, S. and Wu, C. (2002). Comparing the effect of different design of desk with regards to motor accuracy in writing performance of students with cerebral palsy. *Applied Ergonomics,* 34, 141–147.

[5] Wall, M., Van riel, M.P.J.M. and Snijders, C.J. (1991). The effect on sitting posture of a desk with a 10° inclination for reading and writing. *Ergonomics*, 34(5), 575–584.

Emotions as Determinants of Well-Being

Myth or Reality

Sarika Patel[1] and Upinder Dhar[2]

ABSTRACT

The World Health Organisation describes 'well-being' as a "resource for healthy living" and "positive state of health" that is "more than the absence of an illness" and enables people to function well: psychologically, physically, emotionally and socially. Emotional well-being refers to the emotional quality of an individual's everyday experience—the frequency and intensity of experiences of joy, stress, sadness, anger, and affection—that makes one's life pleasant or unpleasant. Emotional well-being is regulated by a number of factors like income, age, work environment, stress, humour, resilience, social interaction and physical health amongst others. It is possible to improve one's emotional well-being through various ways like mindfulness, positive relationships and social support amongst others.

Keywords: Broaden-and-Build Model, Emotional Regulation, PERMA, Problem Solving, Resilience.

INTRODUCTION

Some people think that well-being is being happy and joyous all the time while others prefer to view well-being as a prolonged state of contentment. For others still, well-being is simply about wellness – having good physical and mental health. The term well-being is much more than all these constructs. It encompasses all the ways in which people experience and evaluate their lives positively (Tov, 2018). There are two major approaches to conceptualizing well-being (Ryan & Deci, 2001). The first approach emphasizes on a person's evaluation of his/her own life – both emotionally and cognitively. It has been referred to as hedonic well-being (Diener, 1984), and the second approach includes several concepts that together have been referred to as eudaimonic well-being (Ryan & Deci, 2001). The concept of psychological well-being (Ryff, 1989) is an example of

[1]Research Scholar, Shri Vaishnav Institute of Social Sciences, Humanities and Arts, Shri Vaishnav Vidyapeeth Vishwavidyalaya, Indore, Madhya Pradesh, India. E-mail: sarikap1702@gmail.com

[2]Vice Chancellor, Shri Vaishnav Vidyapeeth Vishwavidyalaya, Indore, Madhya Pradesh, India. E-mail: vc@svvv.edu.in

the eudaimonic well-being tradition which is rooted in the pursuit of goals and activities that are consistent with one's values and identity (McGregor & Little, 1998; Waterman, 1993).

RESEARCH QUESTIONS

This study has the following research questions:

1. How emotions contribute to well-being?
2. Is quality of emotions as important as their experience?
3. How researchers have studied emotions and well-being in the past?
4. How can well-being be improved through emotional regulation?

POSITIVE EMOTIONS AND WELL-BEING

Seligman gave the concept of PERMA, which stands for Positive Emotion, Engagement, Relationships, Meaning and Accomplishment as five measurable elements that makes up well-being (Seligman, 2011). He suggested that positive emotions are important for one's well-being.

Broaden-and-build model of positive emotions provides the foundation for prevention of serious mental health problems like anxiety and depression (Fredrickson, 1998). According to this model, the form and function of positive and negative emotions are distinct and complementary. Negative emotions narrow an individual's momentary thought–action repertoire whereas positive emotions broaden an individual's momentary thought–action repertoire, which in turn can build that individual's enduring personal resources. A range of intervention and coping strategies optimize health and well-being to the extent that they cultivate positive emotions. Cultivated positive emotions not only counteract negative emotions, but also broaden individuals' habitual modes of thinking and build their personal resources for coping. This leads to better health and well-being (Fredrickson, 2000).

On the similar lines, it was reported that approach motivations (direction of behaviour towards positive stimuli like objects, events and possibilities) may influence well-being not only through emotion over time but also through the degree to which people weigh particular emotional experiences in broader judgments of satisfaction (Updegraff, Gable & Taylor, 2004). Macht (2007) studied both individual characteristics and emotional features, and specified five classes of emotion-induced changes of eating. They point to basic functional principles underlying the relations between emotions and biologically based motives: interference, concomitance and regulation. It was found that emotion-induced changes of eating can be a result of interference of eating by emotions, a by-product of emotions, and a consequence of regulatory processes (i.e., emotions may regulate eating, and eating may regulate emotions).

Emotions and Problem Solving

Recent experimental studies show that emotions can have a significant effect on the way people think, decide, and solve problems. In other words, the emotions of an individual have an effect on reasoning and performance of independent tasks. In particular, the participants in a negative emotional state were more likely to deviate from logical norms. The participants in a positive state were also not as good as the neutral group, but this difference was less pronounced. Overall, participants in a neutral emotional state performed best on problem solving and decision-making tasks (Jung *et al.*, 2014).

EMOTIONS AND HEALTH

Emotions-health research has concentrated on primary causation, stating the notion that emotions directly contribute to disease etiology and/or progression. Research has focused on trait or chronic affect with characteristically greater negative affect having been linked to heart disease (Donker, 2000), cancer (Reiche, Morimoto & Nunes, 2006), arthritis (Kiecolt-Glaser *et al.*, 2002), diabetes (Carnethon *et al.*, 2003), and the common cold (Cohen *et al.*, 1998). Greater positive affect has been linked to lower hospital readmission (Middleton & Byrd, 1996), reduced risk of stroke (Ostir *et al.*, 2001), and greater positive affect across the last week predicts lower mortality in both older groups (Ostir *et al.*, 2000) and chronically ill samples (Moskowitz, 2003).

Research suggests that global affective states – feeling good or bad – contribute to unhealthy behaviours such as smoking (Addicott, Gray & Todd, 2009; Perkins *et al.*, 2008), alcohol consumption (Kelly, Masterman & Young 2011; Ostafin & Brooks, 2011), and overeating (Loxton, Dawe & Cahill, 2011). Emotions also contribute to health-related risk perceptions (Peters, Lipkus & Diefenbach, 2006) and health decisions made in response to numeric information (Peters *et al.*, 2009). Less systematic attention has been paid to discrete emotions such as anger, fear, sadness, or disgust (and positive discrete emotions such as gratitude or pride), despite evidence in other domains demonstrating that emotions of the same valence (e.g., anger and fear) can yield dramatically different decisions and behaviours (Lerner & Keltner, 2000; 2001).

Emotions and Memory

Levine and Burgess (2011) examined the effects of happiness, anger, and sadness on participants' memory for different types of information in a narrative. It was found that participants in the positive emotional condition recalled more of the narrative as a whole than did participants in the negative emotional condition. Analyses based on self-reported emotions indicated that happiness had a general facilitative effect on recall, whereas anger and sadness were associated with enhanced recall of information concerning goals and outcomes respectively. These findings indicate that specific emotions differ in their effects on memory and that negative emotions may facilitate selective encoding of functional information.

Emotions and Mental Health

While emotions can be adaptive in many ways, emotions can also be maladaptive (Amstadter, 2008). For example, researchers have noted that while positive emotional experiences and emotional intelligence have a positive effect on mental health, suppression of emotions, inability or difficultly in understanding and communicating emotions (e.g., alexithymia), disposition to exaggerate emotional experiences may have health impairing effect (Pandey and Choubey, 2010).

Difficulties in emotional regulation or emotional deregulation have been found to be associated with a number of mental health problems, such as depression (Gross and Munoz, 1995), anxiety disorders (Campbell-Sills *et al.*, 2006), post-traumatic stress disorder (McDermott *et al.*, 2009), and social dysfunction (Gross and Munoz, 1995). Difficulties in emotional regulation have also been linked with high negative affect and lower level of positive affect and satisfaction with life (Quoidbach *et al.*, 2010).

Emotions and Resilience

Positive emotions are responsible for increases in both resilience and life satisfaction whereas negative emotions do not. Positive emotions also mediated the relation between baseline and final resilience, but life satisfaction did not. This suggests that it is the momentary experience of positive emotions, and not more general positive evaluations of one's life, that form the link between happiness and desirable life outcomes. Change in resilience mediated the relation between positive emotions and increased life satisfaction, suggesting that happy people become more satisfied not simply because they feel better but because they develop resources for living well (Cohn *et al.*, 2011).

Emotions and Cognition

Researches in the past have found that emotions are important in education. Moreover, emotional well-being should also be regarded as an educational goal because of its multifaceted benefits for students. Emotions have an impact on students' attention, influence their motivation to learn, modify the choice of learning strategies, and affect their self-regulation of learning. Furthermore, emotions are part of students' identity, and they affect personality development, psychological and physical health. Emotions are important because of their influence on learning and development (Pekrun, 2014).

Studies on affective well-being have shown the beneficial role of positive emotions on cognitive processing and the harmful role of negative emotions on coping, stress and health status. In today's world, although modern technology has saved mankind from physical damages like injuries, infections, accidents, but one comes across many emotionally demanding situations in all fields of life, such as high-level competition, unemployment, deadlines, high expectations at the workplace, adjustments in families, dealing with difficult personalities, etc. causing heightened anxiety and stress. It has been found that integrated yoga practices can reduce the negative affect and increase the positive affect within one week in healthy young adults (Narasimhan, Nagarathna & Nagendra, 2011). Emotions can be regulated well by means of practicing yoga.

EMOTIONS IN INDIAN CONTEXT

In the Indian philosophical texts, detailed descriptions of emotions are not available nor are dealt with as a separate concept. This view of emotions lays emphasis on desires as the root cause of emotional upheavals. Emotions are viewed as springs of action and are bipolar in nature. According to Patanjali's Yoga Shastra, suffering is due to ignorance about one's true "self" (avidya). Hence, suffering or dukha arises from within and not from the outside world. Bhagvadgita traces all emotional experiences to the gunas i.e., sattva, rajas, and tamas. Works of Bharathmuni have contributed to the understanding of emotional experiences. These views underline the recommended path for self-transformation. Regulating emotions, both emotional experience and emotional expression, is an integral part of the recommended "principles of living" (Ramaprasad, 2013).

This study suggests that the key factors of good health like positive thinking of love, courage and optimism, social support, self-esteem, purpose in life, positive emotions, spirituality, eating right food and a positive mental attitude, not only improve the quality of people's lives but also heal their bodies from psychological and physiological illnesses (Singh and Pareek, 2007). Humans are emotional beings. People feel happy when they are with their loved ones; sad if they lose someone very close; guilty if they hurt someone unknowingly; and angry if things do not happen as planned. Hence there is never a moment when people do not emote. What are humans without emotions? They would be mere machines working like a robot. Emotions colour our life experiences. Emotions inform people as to who they are, and their relationship with others around them, and how they must act. Emotions give meaning to our existence and the events that people experience. Without emotions, these experiences would be mere pieces of information. Taking a holistic view of health, emotions are integrated with a particular emphasis on Indian tradition of thought, the theory of Karma. Thus, intention is to stimulate a novel way to look at feelings and improve the understanding of emotional responses (Khosla, 2010).

Health related challenges are the plausible consequence of maladaptive emotional regulation that leads to emotional and behavioural disturbances amongst people. Studies reveal that health and emotional regulation of well-being may influence help seeking behaviour, symptoms report, medical adherence and somatic health of the people. The way people regulate their health is shaped by bio-psychosocial and cultural processes. Gender differences have also been reported in emotional regulation strategies as studies indicate that women are comparatively better than men in attending, valuing, and expressing emotions in daily life. Other findings have reported that styles of emotional expression are related with health as individuals who repress their emotional symptoms exhibit poor health outcomes. While the Individuals, who feel emotions with greater intensity, differentiate among them and regulate them properly, show better health outcomes in their health-related issues (Verma & Kumar, 2016).

Findings of this study indicate that some specific types of emotional deficits such as difficulties in identifying feelings, lack of emotional clarity and limited access to emotion regulation strategies were relatively more important in predicting the health status and well-being of an individual as compared to other emotional difficulties. Also, the findings imply that emotional regulation difficulties and alexithymia in general is associated with impaired mental health and lower level of happiness and life satisfaction, i.e., subjective well-being (Saxena, Dubey and Pandey, 2011). Emotional regulation is an important component of human development and well-being. Research findings suggest that a habitual emotional state predicts long-term health and life expectancy, and that successful emotion regulation is vital to mental health as well as physical well-being. Studies also indicate that individuals regulate their emotions in wide variety of ways some of which are healthier than others (Singh & Mishra, 2011).

CONCLUSION

The studies included in this paper have shown that how emotions are crucial to one's well-being. People experience a number of emotions and feelings on day-to-day basis. These emotions affect the way people think, decide, solve problems, cope with stressors, read, write, maintain positive relationships and bounce back from difficulties. Emotions can improve and/or deteriorate one's bio-psychosocial health. Number of researchers have proved that emotional quotient is more important than Intelligence Quotient (IQ) to excel at work, social settings and life in general. It has also been established that emotional well-being is important for happiness and life satisfaction, the ultimate aims of human lives. Thus, it is imperative that more attention is paid to how one feels, cultivates more positive emotions and has a good mental, physical and social health.

IMPLICATIONS

Emotions should be given more importance and training for emotional regulation and processing should be given to every individual as emotions are very crucial to one's well-being. There are several methods like yoga and meditation, positive relationships, resilience training and healthy coping strategies which have been proved very useful in regulating one's emotions and eventually improving one's well-being. The paper has covered a limited time period. Further work can be done on the same lines with more literature. Further studies can also focus on other aspects of well-being, such as the interventions related to emotional well-being.

REFERENCES

[1] Amstadter, A. (2008). Emotion Regulation and Anxiety Disorders. Journal of Anxiety Disorders, 22(2), 211–221. doi: 10.1016/j.janxdis.2007.02.004.

[2] Campbell-Sills, L., Barlow, D.H., Brown, T.A. and Hofmann, S.G. (2006). Acceptability and Suppression of Negative Emotion in Anxiety and Mood Disorders. Emotion, 6(4), 587–595.

[3] Cohn, M.A., Fredrickson, B.L., Brown, S.L., Mikels, J.A. and Conway, A.M. (2009). Happiness unpacked: Positive emotions increase life satisfaction by building resilience. *Emotion, 9*(3), 361–368. https://doi.org/10.1037/a0015952.

[4] Diener, E. (1984). Subjective well-being. *Psychological Bulletin, 95*, 542–575.

[5] Fredrickson, B.L. (2000). Cultivating positive emotions to optimize health and well-being. Prevention & Treatment, 3(1), Article 1. https://doi.org/10.1037/1522-3736.3.1.31a.

[6] Gross, J.J. and Munoz, R.F. (1995). Emotion Regulation and Mental Health. *Clinical Psychology: Science and Practice*, 2(2), 151–164.

[7] Jung, N., Wranke, C., Hamburger, K. and Knauff, M. (2014). How emotions affect logical reasoning: evidence from experiments with mood-manipulated participants, spider phobics, and people with exam anxiety. Frontiers in psychology, 5, 570.

[8] Khosla, M. (2010). Understanding emotions from an Indian perspective: Implications for well-being. Emotional expression: The brain and the face.

[9] Levine, L.J. and Burgess, S.L. (1997). Beyond general arousal: Effects of specific emotions on memory. Social Cognition, 15(3), 157–181.

[10] Macht, M. (2008). How emotions affect eating: a five-way model. *Appetite, 50*(1), 1–11.

[11] McDermott, M., Tull, M., Gratz, K., Daughters, S. and Lejuez, C. (2009). The Role of Anxiety Sensitivity and Difficulties in Emotion Regulation in Posttraumatic Stress Disorder among crack/cocaine Dependent Patients in Residential Substance Abuse Treatment. Journal of Anxiety Disorders, 23(5), 591–599. doi:10.1016/j. janxdis.2009.01.006.

[12] McGregor, I. and Little, B.R. (1998). Personal projects, happiness, and meaning: On doing well and being yourself. Journal of Personality and Social Psychology, 74, 494–512.

[13] Pandey, R. and Choubey, A.K. (2010). Emotion and health: An overview. SIS Journal of Projective Psychology & Mental Health, 17(2).

[14] Pekrun, R. (2014). Emotions and learning. Educational practices series, 24(1), 1–31.

[15] Quoidbach, J., Berry, E.V., Hansenne, M. and Mikolajczak, M. (2010). Positive Emotion Regulation and Well-being: Comparing the impact of eight savoring and dampening strategies. Personality and Individual Differences, 49(5), 368–373.

[16] Ramaprasad, D. (2013). Emotions: an Indian perspective. *Indian journal of psychiatry, 55*(Suppl 2), S153.

[17] Ryan, R.M. and Deci, E.L. (2001). On happiness and human potentials: A review of research on hedonic and eudaimonic well-being. *Annual Review of Psychology, 52*, 141–166.

[18] Ryff, C.D. (1989). Happiness is everything, or is it? Explorations on the meaning of psychological wellbeing. *Journal of Personality and Social Psychology, 57*, 1069–1081.

[19] Saxena, P., Dubey, A. and Pandey, R. (2011). Role of Emotion Regulation Difficulties in Predicting Mental Health and Well-being. *SIS Journal of Projective Psychology & Mental Health, 18*(2).

[20] Singh, S. and Mishra, R.C. (2011). Emotion regulation strategies and their implications for well-being. *Social Science International*, 27(2), 179.

[21] Singh, S. and Pareek, R. (2007). Positive predictors of health. *Journal of Indian Health Psychology, 2(1),* 10–19.

[22] Tov, W. (2018). Well-being concepts and components. In E. Diener, S. Oishi and L. Tay (Eds.), Handbook of well-being. Salt Lake City, UT: DEF Publishers. DOI: nobascholar.com

[23] Updegraff, J.A., Gable, S.L. and Taylor, S.E. (2004). What Makes Experiences Satisfying? The Interaction of Approach-Avoidance Motivations and Emotions in Well-Being. Journal of Personality and Social Psychology, 86(3), 496–504. https://doi.org/10.1037/0022-3514.86.3.496

[24] Verma, N.K. and Kumar, S. (2016). Role of health and emotions in promotion of well-being: Current thoughts in Indian perspective. Indian Journal of Health & Wellbeing, 7(8).

[25] Waterman, A.S. (1993). Two conceptions of happiness: Contrasts of personal expressiveness (eudaimonia) and hedonic enjoyment. *Journal of Personality and Social Psychology, 64*, 678–691.

Event Sentence Classification Model based on Event Trigger and Argument from News Articles

Nikhil Chaturvedi[1] **and Jigyasu Dubey**[2]

ABSTRACT

Events play an important role in understanding the phenomena happening in the real world. Event is a very common understanding relations between objects, place, actions, and people. Events centered modeling involves various aspects of an event along with the semantic representation of event facts. For the event sentence classification, a model involving the event identification, event trigger identification and event argument extraction has been proposed. The event sentence classification model has been divided into two steps: event trigger and argument identification and second step is the event sentence classification using training corpus. In the process of evaluation of model, the manually tagged news articles are compared with the automatically identified event sentence by the model.

Keywords: Arguments, Event Extraction, Event Trigger, News Articles.

INTRODUCTION

In new era of online published news articles, hundreds of articles are published every day by the popular news agencies like CNN, BBC, etc. The amount of online published news articles increases day by day, so the reader needs to find news articles related to specific event type of his or her interest. In general, a reader finds news articles of his or her interest, by filtering out the desired from headlines and teaser by scanning various sources of news articles. People always want to extract and comprehend a huge amount of information in less time. A reader not willing to read through a collection of news articles needs a representation of the news in a compact form describing the event type briefly. The amount of unstructured data enormously increases every day, so a model is required to automatically identify the several types of knowledge from the data

[1]Research Scholar, Shri Vaishnav Institute of Information Technology, Shri Vaishnav Vidyapeeth Vishwavidyalaya, Indore, Madhya Pradesh, India. E-mail: nikhilcse2015@gmail.com

[2]Professor and Head, Shri Vaishnav Institute of Information Technology, Shri Vaishnav Vidyapeeth Vishwavidyalaya, Indore, Madhya Pradesh, India. E-mail: jigyasudube@yahoo.co.in

(Chen *et al.*, 2015). A recurrent difficulty in processing large amount of data in a systematic manner is because most of the data are in unstructured form, e.g., data written in human understandable language.

This information is in the form of online broadcast news, blogs to comments and social communication data (Haibo *et al.*, 2019). Thus, it may also consist of various types of unrelated data or noisy discussions. Event extraction is the task to identify the key terms related to the specific events from unstructured data. Event is defined as a specific occurrence involving participants. It is something that can be explained as a change of state, so in various fields it is classified in terms of events. Formally, the event extraction is to automatically identify events in collection of text and to find detailed information about them. Event extraction from texts could be beneficial to various domains. In the online published news articles application, the ability to identify events may enhance the performance of specific news recommendation. The online published news articles may be selected more efficiently based on the specific extracted events and user preferences (Hamborg *et al.*, 2018).

Event extraction and two categorization of online published news articles have many applications such as government policymakers uses the information related to significant international incidents tracking to make better decisions regarding policies. In cyber security domain, ongoing cyber-attack incidents help to monitor. Another most valuable application of event extraction is in stock market which is extremely sensitive to the breaking news for deciding the stock trading planning such as time, price, and volume. Events are extracted from text at sentence level. Some past studies have shown extraction of events using relations, entity, and article level (Sarma *et al.*, 2018). Sentence level event extraction is very difficult task, as the event trigger and argument detection heavily depends on it (Yang, 2019). In the news articles it can be challenging to identify the same event that have different form, either present in single or multiword units. Event extraction comes from the unstructured data, where many non-event examples are also present, thereby making event and argument identification a challenging task (Yarrabelly *et al.*, 2019).

For such essential characteristics of online published news articles, documents require more semantic information processing, for building a larger scale news semantic knowledge base (Tarakeswara *et al.*, 2020). In the current situation this type of task is performed manually by the news article analysts and online digital news editors. They collect, interpret, and present news from several sources and require a system that can manage events automatically for the use in news application where new events and existing ones move on smoothly. Event situation is explained using event arguments. Event arguments provide a view of multidimensional understanding of the event description. Event extraction system is heavily dependent on the language therefore event extraction system has to be developed by various languages. Taking some language related system into consideration, a various event extraction model has been developed in various languages English (Alsaedi *et al.*, 2015), Chinese (Chen *et al.*, 2012), Turkish (Can *et al.*, 2009) and others.

REVIEW OF LITERATURE

Chen *et al.* (2015), proposed a dynamic multipooling conventional neural network model (DMCNN), according to event trigger and argument to receive more information. Tillo *et al.* (2015) proposed a web toolkit for event extraction which automatically evaluated the suggestions and explorations and showing to the user how the events are related to one another. Grishman *et al.* (2016) described two stage training model based on neural networks that is effectively used for the transformation of knowledge from the event types to the goal event types. They also investigated the effectiveness of the method compared with other state of art methods. This method for extraction of knowledge may be more efficient in the use of real-time dataset for description. Yagung *et al.* (2017) presented a timeline that shows significant events about query word based on the corpus, which can inform about the user's interest.

In the procedure to cluster sentences by their dates, they consider the value of no precise date sentence. Because it is now one date vector for one date, they can only extract the most important thing in each date vector. It is possible that several things with same type happened on same date. Liu *et al.* (2017) presented the extraction of argument information explicitly for event detection using supervised mechanisms. In this model, researchers used systematic approach for the investigation of proposed model under the guidance of various attention strategies. Keshavarz *et al.* (2017) proposed a novel probabilistic inference framework for complex video event recognition using supervised action concepts. In this method author attempted to model for the conditional relationships between complex events and exhaustive set of intermediate concepts by constraining dependencies to pairwise joint distributions while avoiding the need of manually reencoding new graph structures as the number of concepts increases.

Namazifar *et al.* (2018) suggested a Named Entity Recognition (NER) method aimed at locating and classifying named entities in the text. Proposed NER method applies to find Named entity from Tweets. Researcher suggests that the NER is also used for the online news article texts. Jacoab *et al.* (2018) presented a dataset and categorization model for company specific economic event detection from the economic English news articles. So that in this model room for the improvements like more annotated data and other resources are required to further ambiguous event sentences. Hamborg *et al.* (2018) proposed open-source system that retrieves answers in the question forms of what, when, who, where, why and how to describe online published news articles information in main events.

Yang *et al.* (2018) presented to obtain event profiles event extraction automatic framework using unsupervised learning method. Their proposed framework was based on event profiling. For implementation of work, lots of challenges like finding out event types in given event initially unknown for users, lack of event types of schemas in previously defined models, occur. So that event profiling to their event types is still demanding in the field of information extraction from text. Ding *et al.* (2019) presented a method based on word and contexts for Human Needs categorization. They developed a method for assigning words to related concept in between of categorization process. Researchers are creating supervised model for labelling the words into their semantic concepts for human need information categorization. They use 10 types of semantic concepts in their work.

Walkar *et al.* (2018) presented a new model that is based on GRU that combines the temporal structure information and syntactic information in that mechanism. The result showed that it is competitive with other state of art methods like neural network architectures using empirical evaluations under the split of ACE dataset. They suggested that event can be categorized using their types. Yang *et al.* (2019) proposed an efficient collapsed Gibbs sampling approach to infer the truths of news and the users' credibility without any labelled data. In this model, researcher showed that the experimental results on two different data sets outperforms the other compared methods. Wang *et al.* (2019) proposed a new approach based on adversarial training for the extraction of the structured presentation of events from the online text. In this proposed model, researcher compared the model with the different state of arts method and showed that AEM improved the extraction performance especially for long text. Future work can explore incorporating external knowledge (e.g., word relatedness contained in word embedding) into the learning framework for event extraction.

PROPOSED MODEL

Events are very dynamic that perform an important role to understand the real-world problems, which are mostly dependent upon what, who, when and where type of questions. Without such types of question, we cannot talk about the events, because event always require the relations between the places, actions, people, and objects. So that the event centered modelling always requires the various aspects of an event and event participation entities. Accordingly, some assumptions to presents event facts, some parameters used for defining the events using four dimensions are Event E = {E, T, P, L}.where E is an real world phenomenon event entities, T is the specific time when event occurs, P is all participant involved in the event and L for the location where the event happened or published in online news articles. The proposed event modelling architecture has four main components which are pre-processing, event trigger identification, event argument extraction, and sentence classification. Figure 1 shows the architecture Event Sentence Classification Model based on event trigger and argument from news articles.

Pre-Processing

It is preliminary task for event sentence extraction process. The main role of the pre-processing is to get ready the news article for event sentence extraction processes. This pre-processing module accepts the unstructured raw data of natural language texts as input for processing. In this module the process is divided into two parts in the first part tokenization and in the other normalization is performed. In the raw text of news article firstly we find the where the sentence starts and end after that we find the whitespace, identify word and punctuation marks in the text. Sentence and word tokenization is performed using punctuation marks and white space. In the process of normalization task, mapping of the pre-processed text or expression to their standard format for measurements is done. In the normalization process two types of normalization for English news articles one is character and other is temporal expression normalization was done. In Character normalization the different characters having the same pronunciation into a single character were normalized.

In the present model, character normalizer finds the similar character and replaces it with the normalize character issue related to the ambiguity in the text and the role based temporal expression normalizer was used. Finally, after the pre-processing segmented news article into sentences and tokens, normalized that makes news article ready for other tasks were found.

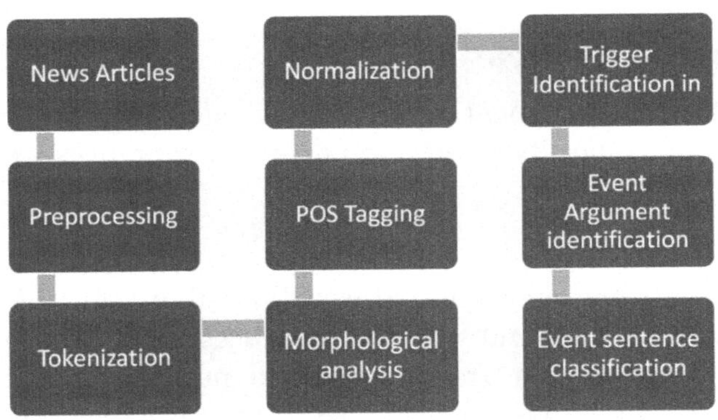

Fig. 1: The Architecture Event Sentence Classification Model based on Event Trigger and Argument from News Articles.

Trigger Identification

Event trigger word is the identity of the event, which is presented in all essential features for finding of the event mention. For finding such types of word wide range of analysis of textual features is required to identify the best trigger to the task of event extraction. In this module of present model, all event trigger word from the available training corpus containing news article of several domains were collected. In the news article every single word cannot be the candidate of event trigger so that the best possible method for event trigger identification method was used. Firstly, noun and verbs are the trigger words for the event so that help of POS tagger the noun and verbs present in the news article were found and after that the trigger word in the sentences present in the news articles were found. In the trigger identification method for selecting only very informative triggers POS tagger and morphological analysis was used.

Event Argument Extraction

In the module of Event argument extraction involves entities during all process of event. Event argument extraction module is accountable for the identification of the basic information like extent, temporal information, participants, and spatial elements. Each element of the argument extraction module can be explained by several ways and value after the processing of the news articles. Normally event arguments are explained by using some facts of what, where, when and who of news articles. In the process of event extraction event argument extraction is most important task for the selection of the various entities mentioned in the news articles, but only few of the entities and value fall within the scope of the event extraction. To select such types of elements or argument in the news articles entity recognizer and other local features like potential trigger and path form the head word of the event trigger has been used.

Event Sentence Classification

In this module for the classification of news article sentence Maximum Entropy classifier algorithm to classify event and non-event sentences were used.

CONCLUSION

Events are essential characteristic real-world possibility for the human. Event is made of what, when, where and who that produces up of event trigger and event arguments. Important events are published in news article like any mundane events. In the present paper an event sentence classification model based on event trigger and argument from news articles has been proposed. In this proposed modelling task involves event identification, event trigger identification and event argument extraction from the news article datasets. Event trigger identification and event argument extraction is done by extensive analysis of unstructured news article data. To identify the event mention, ACE2005 Proposed event triggering words for various news articles were used. Named entity recognizer (NER) and other local features are also used to identify event argument. For the classification of news article sentence we used Maximum Entropy classifier algorithm to classify event and non-event sentences.

REFERENCES

[1] Alsaedi, N. and Burnap, P. (2015). "Feature Extraction and Analysis for Identifying Disruptive Events from Social Media", *IEEE/ACM International Conference on Advances in Social Networks Analysis and Mining,* Cardiff School of Computer Science & Informatics Cardiff University, Cardiff, UK.

[2] Can, F., *et al.* (2009). "New Event Detection and Topic Tracking in Turkish", *Journal of the American Society for Information Science and Technology.*

[3] Chen, Y., Xu, L., liu, K., Zeng, D. and Zhao, J. (2015). Event Extraction via dynamic multi polling convolutional neural networks. In *ACL-IJCNLP2*, 167–176.

[4] Chen, Z. and Heng, J. (2016) "Language Specific Issue and Feature Exploration in Chinese Event Extraction", 365 Fifth Avenue, New York, NY 10016, USA.

[5] Ding, H. and Riloff, E. (2018). Human needs categorization of Affective events using labelled and unlabelled data. In: *Proceedings of the 2018 Conference of the North American Chapter of the Association for Computational Linguistics: Human Language Technologies.*

[6] Ding, H., Jiang, T. and Riloff, E. (2018). Why is an even affective? Classifying affective events based on human needs. In: *The AAAI-18 Workshop on Affective Content Analysis.*

[7] Haibo, D., Ellen, R. and Zhe, F. (2019). Improving human needs categorization of events with semantic classification. *Proceeding of Eighth Joint Conference on Lexical and Computation Semantics by ACL.*

[8] Hamborg, F. and Gipps, B. (2018). Automated Identification of media bias in news articles: an interdisciplinary literature review. *International Journal on Digital Library*, Springer 16 November.

[9] Hamborg, F., Breitinger, C., Schboutz, M. and Gipps, B. (2018). Extraction of main event descriptor from news articles by answering the journalist five W and one H questions. *FCDL'18,* June 3–7, Fort Worth, TX, USA.

[10] Hamborg, F., Breitinger, C., Schboutz, M. and Gipps, B. (2018). Giveme5W: main event retrieval from news articles by answering the journalist five W. I Conference in Sheffield, UK.

[11] Jacobs, G. and Lefever, E. (2018). Economic Event detection in company specific news text. In the proceedings of first workshop on Economic and NLP, Melbourne Australia ACL 1–10.

[12 Kamel, M. and Keyvani, N. (2018). Sentiment content analysis and knowledge extraction from news articles. Association of computer linguistics conference, pp. 13–18.

[13] Liu, S. *et al.* (2017). Exploiting argument information to improve event detection via supervised attention mechanisms. Proceedings of the 55th Annual Meeting of the Association for Computational Linguistics 1.

[14] Michael, T. and Akbik, A. (2015). A web toolkit for exploratory relation Extraction. In ACL- IJCNLB.

[15] Namazifar, M. (2017). Named entity sequence classification. arXiv preprint arXiv:1712.

[16] Nassem, M., Chon, b., and Shamail, S. (2019). "Discovering Events from Social Media for Emergency Planning." Association for Computing Machinery, Dubai, UAE.

[17] Nguyen, T.H., Fu, F., Cho, K. and Grishma, R. (2016). A two-stage approach for extending event detection to new types via neural network. In WRepL4NLP, 158–165.

[18] Rao, B.T., Murty, R., *et al.* (2020). A comparative study on effective approaches for unsupervised statistical machine translation. *International conference and published the proceedings in AISC Springer conference* (1076), 895–905.

[19] Sarma, P., Liang, Y. and Sethares, W. (2018). Domain adapted word embedding is for improved sentiment classification. In: *Proceedings of the ACL Workshop on Deep Learning Approaches for Low Resource NLP.*

[20] Singh, S. (2018). Natural language processing for information extraction. *Association of Computer Linguistics Conference*, pp. 13–18.

[21] Somayeh, K., Imran, S. and George, A. (2017). Exploiting probabilistic relationships between action concepts for complex event classification. *IEEE Transaction and Information Theory.*

[22] Suchanek, F., *et al.* (2007). A core of semantic knowledge. In: Proceedings of the 16th international conference on World Wide Web ACM, 697–706.

[23] Walker, J., Tadepalli,P. and Fern, X. (2018). Event detection with neural network: a rigorous empirical evaluation. arXiv preprint arXiv:1808.

[24] Wang, R., Zhou, D. and He, Y. (2019). Open event extraction from online text using a generative advisable network. *Proceedings of 33rd AAAI Conferences on Artificial Intelligence.*

[25] Wei, W. and Dongyan, Z. (2012). "Chinese News Event 5W1H Semantic Elements Extraction for Event Ontology Population", *International World Wide Web Conference Committee (IW3C2),* WWW 2012 Companion, April 16–20, 2012, Lyon, France.

[26] Wu, Y., Sun, H. and Yan, C. (2017). An event timeline extraction method based on news corpus. *International Conference on Big Data Analytics.*

[27] Yang, Q., *et al.* (2018). Open schema event profiling for massive news corpus." *CIKM'18*, Torina, Italy.

[28] Yang, S., Shu, K., Wang, S., Gu, R. and Liu, H. (2019). Unsupervised fake news detection on social media: A generative approach. *Proceedings of 33rd AAAI Conferences on Artificial Intelligence.*

[29] Yarrabelly, N. and Karlapalem, K. (2018). "Extracting Predictive Statements with their Scope from News Articles." *Proceedings of 12th International AAAI Conference on Web and Social Media.*

Folk Elements in the Manuscript Paintings of Bal-Gopal-Stuti

Ruchi Agrawal[1] and Anu Ukande[2]

ABSTRACT

Folk arts are elementary paintings done on walls and floor and uses a very basic graphic vocabulary like lines, circles, squares and triangles. These motifs or elements come from the nature around them, which is a result of man's close association with his immediate surroundings India has always been a subcontinent that portrayed traditional and cultural spirit through its multi-dimensional, conventional arts. Folk art expresses cultural identity by conveying shared community values and aesthetics. Every region has its own unique style and pattern of the primitive or folk art. These native arts of India are ethnic and simple and are distinctive enough to say a lot about its rich heritage. This paper focuses on the folk art of West Bengal called Alpana vis-à-vis the painted folios of Bal-Gopal-Stuti (Hymn of praise of child and youthful Krishna), which is a Sanskrit anthology of praises for Lord Krishna, credited to the saint-poet Bilvamangala, who lived around twelfth or thirteenth century. The manuscript has paintings and text both which hold aesthetics in visual as well as pictorial representations. The verses or the strotras have the inspirations from the epics like Bhagavata Puranas, Bhagavat Gita or other different Puranas. There are at least twelve surviving manuscripts of Bal-Gopal-Stuti made during 200 years from fifteenth-sixteenth century.

The present paper aims to study the folk motifs and elements in the painted folio of miniature manuscript of Bal-Gopal-Stuti (Hymn of praise of child and youthful Krishna). It's a Sanskrit anthology of praises for Lord Krishna, which is credited to the saint-poet Bilvamangala, who lived around twelfth- thirteenth century.

Keywords: Bal-Gopal-Stuti, Bhagavata, Folk Art, Manuscript Painting, Puranas.

INTRODUCTION

Indian culture showcases a rich folk tradition. Folk arts not only comprise of visual arts like painting but also the performing arts like folk-music, folklore,

1 Research Scholar, Shri Vaishnav Institute of Fine Arts, Shri Vaishnav Vidyapeeth Vishwavidyalaya, Indore, Madhya Pradesh, India. E-mail: ruchi.pankaj104@gmail.com

2 Assistant Professor, Shri Vaishnav Institute of Fine Arts, Shri Vaishnav Vidyapeeth Vishwavidyalaya, Indore, Madhya Pradesh, India. E-mail: anuukande1980@gmail.com

drama, and dance. Folk arts play an important role in handing down the rituals and customs with their religious knowledge, beliefs, faiths, and rites. It is in the form of inherited traditions which frame the identity of each society. These established beliefs and practices carry aesthetic relish along with maintaining their ethnic nature and originality. Folk people express themselves in vivid styles through the paintings, which was the only means of transmission and inculcation of the culture through folk lore to a populace those who were not familiar with the written word (Pandey, 2020).

Folk paintings are mostly pictorial representations and there is a variation in subjects from religion to nature and day-to-day activities. The folk arts are generally related to their rituals and traditions which are made during/for cultural festivals, puja ceremonies, or for the ceremonies connected with harvest, marriage, birth, and death. The myths of a particular region form the plot of these art styles. Painted narratives on various surfaces such as floor, cloth, wood, paper, wall, palm leaf etc. are important sources of knowledge about deities and historical heroes, local mythology, epics, folktales, and customs. The designs are generally freehand drawings, mostly done by the women of the households. Stylised depiction of the sun, moon, birds, flowers, zodiacal signs, trees and nature inspired motifs form part of the design.

Along with bright and vibrant colours, folk arts have an appealing charm. There is a strong decorative orientation in all the motifs and patterns depicted in folk arts. The decorative designs draw heavily on geometrical shapes and simplification of complicated shapes of plants and animals. The prevalence of flora and fauna themes shows the significance of nature in folk life. The folk arts in textile, pottery, carving, costumes etc reflects the culture of that region. These paintings usually have vibrant bright colour schemes painted using various natural ground substances. In the days of yore, they were made with natural colours and dyes made of soil, mud, cow dung, leaves, charcoal giving it a sense of antiquity and vintage nostalgia.

Alpana is folk art of West Bengal done on floors with traditionally two colours-Indian-red and white. For the red colour *geru* is used, and white is acquired by mixing *rice powder* with water. Both colors used in alpana come from natural sources and alpana is painted on a base of cow dung, water and mud. Eventually in history, scripts and writing developed and many manuscripts started being written. One such manuscript is the Bal-Gopal-Stuti. This manuscript had miniature paintings as well as shlokas praising Lord Krishna. '*Miniature Painting*' as the name suggests is painting of an exceedingly small scale, which is also a narration of a story, it depicts a running tale and not just a moment in time. Miniature paintings of the Bal-Gopal-Stuti continue like a series of the storyline theme. They stand equivalent to expressing the tale just like in literature. Miniature paintings seem to have the ability to give the viewer a complete visual depiction of the text of the manuscript. They seem to portray the mythological stories as something real and true happenings.

Miniature paintings came into existence during the seventh and eighth centuries. They were used as a means of trade, since they could be carried easily because of their compact size and easy handling. Text donations known as *Shastradaan* among followers of Buddhism & Jainism became popular which led to more commissioning of miniature works. These texts were illustrated with gold which were like financial assets and could fetch loan on mortgage. Thus, the miniature paintings replaced the wall paintings due to their usefulness and gave birth to new art styles of India.

By the twelfth century, Muslims had invaded India. Temples were destroyed and images defaced. Their destruction of public monuments endangered the temple frescoes and murals, and hardly anything but literary descriptions of wall paintings survived in northern India. The images painted in books illustrating sacred texts, however, they were safe. Sheltered in private libraries, they were preserved for centuries. In the retreat of quiet and private libraries they could be meditated upon without fear. Because the pages of the earliest books were made of palm leaves, the illustrations had to be small (Kramrisch, 1986).

Miniatures depict life of the era in which they were painted. They were used as illustrations in books to exemplify the scene talked about in the paragraph. One such manuscript painting is the Bal-Gopal-Stuti which is a portrayal of Krishna-Lila. A poet-saint known as Bilvamangala is supposed to have lived during twelfth-thirteenth century. He wandered all around India and sang praises in the form of small *strotras*. This Sanskrit anthology for Lord Krishna must have gained popularity and therefore a need to have it in a manuscript form must have led to paintings. The Fifteenth and sixteenth centuries were a period of expansion and experimentation in manuscript illustration in the western Indian style and there are at least twelve surviving manuscripts of the Bal-Gopal-Stuti created in those 200 years (Dahejia, 2019).

ANALYSING THE PAINTED FOLIOS IN LIGHT OF FOLK ELEMENTS

Taking up a folio from the Bal-Gopal-Stuti, where Krishna is playing flute on the banks of Yamuna (Figure 1). In this image, Krishna has been painted along with three *gopi's*. The other male figure with folded hands is supposed to be the poet Bilvamangala who is describing the scene of delight seeing Krishna playing the enchanting flute. The *gopis* are thrilled to have Krishna around them and the *Shringara rasa* is the prominent sentiment emitted. At the bottom of the painting sacred Yamuna river is shown with curved lines and fishes in a single row gives a feel of continuity and harmony. These are the patterns we can also find in the folk tradition of alpana (Figure 4). A sense of rhythm is also felt by viewing the sky which is shown by three wavy lines in red, blue and white at the top of painting indicating a monsoon sky.

Fig. 1: Credit: Museum of Fine Arts, Boston
Medium / Technique: Opaque Watercolour on paper
Dimension: Each Page: 10.5 × 23.6 cm (4 1/8 × 9 5/16 in)

Fig. 2: Credit: Wellcome Library, London
Medium / Technique: Opaque Watercolour on paper
Dimension: Each Page: 10.5 × 23.6 cm (4 1/8 × 9 5/16 in)

Fig. 3: *Picture Credit:* Wellcome Library, London
Medium / Technique: Opaque Watercolour on paper
Dimension: Each Page: 10.5 × 23.6 cm (4 1/8 × 9 5/16 in)

Krishna is wearing a *pitamber* (yellow) *dhoti, patterned lower garment* in double combination which has motifs of flowers, curved, zig-zag lines, dots, circles and semi-circles. These motifs are still seen in contemporary folk designs. The tree is shown in a balloon like shape which is filled with leaves. This geometrical representation resembles the basic geometrical motifs found in folk arts especially alpana (Figure 5). The tree trunk also is drawn with criss-cross lines. These infills or *bharti* are used to cover the spaces between lines in almost all folk arts, with slight variations. The alpana image show the curved lines used in repetitive pattern. The patterns seen in *Alpana* (Figure 6) resembles the patterns in the folios of Bal-Gopal-Stuti. The infills in the shapes are mostly textural rather psychological.

Fig. 4: Alpana with Geometrical Shapes, Lines, Footprints of Goddess

Fig. 5: Alpana with Floral Motifs

Fig. 6: Alpana Design with Curved Lines, Dots and Circle

Another painted folio from Bal-Gopal-Stuti (Figure 2) showing child Krishna suckling on the foot thumb, lying hidden beneath the leaves has many elements inspired by alpana patterns. A huge tree has been painted on this folio. The tree is surrounded with water and tortoise. Curved lines are drawn in rhythmic pattern to show the ocean which feels like the waves. Bold contour lines are used for the tree which dominates the space and the branches are shown with thin black lines which are filled with dash type of small lines. These kind of lines are still seen in the folk arts. The chequered lines on the tortoise is a common feature of folk arts.

The verses is as follows:

"The moon's soft rays fill the world with love, but force lotuses to close.

This is utterly unbecoming! Contemplating this fact, the wielder of maya.

Lying on a banyan leaf bed,sesSB Joins his own lotus-foot to his moon-face.

As if uniting them.

His sweet and serene smile.

Makes his face so endearing that it repeatedly and joyously sages' minds.

To become lotuses!

Braja's women, whose eyes are languid with drunkenness.

Also relish this lotus-eyed one. When will I ever gaze at another child like him? (Dahejia, 2019). Another folio from Bal-Gopal-Stuti (Figure 3) depicts Krishna lying with Radha on a swing and two other gopis are standing on the sides using *chavar* as a fan.

The swing is still a common-place item of household furniture even today in Gujrat. There is a toran on the swing over them. It has the familiar folk element of flowers, chequered lines and bold multi-coloured lines. The *paan* shaped *toran* is still made in the folk paintings showing the entrance of the household, and is hung on the occasions of any pious rituals, festivals etc.

The spread on the swing has curved lines like a basket-weave or also looks like scales of a fish, pattern reflected in Alpana patterns as well (Figure 5). The furniture is shown with elaborated carvings which is still a craft practised in Gujrat. The drapes of the gopis and Radha also have textile patterns, showing the rich tradition of textile industry of western India. The print is shown with variety of dots and lines which provides diversity in the dresses and reflects alpana motifs (Figure 4).

ELEMENTS FOUND IN IMAGES OF BAL-GOPAL-STUTI AS WELL AS ALPANA

Floral Designs: In the images of Bal-Gopal-Stuti, floral designs can be seen in the dhoti worn by Krishna, on the cloth draped over the swing. The pedestal also has the floral design. The images of Alpana above has motifs of lotus. These are among the popular patterns used, which symbolises happiness and prosperity. They also denotes change in seasons and fertility. Leaves of various trees are considered auspicious which is used as *toran* and in *kalash*. Lotus flower is also drawn for Laxmi pooja or the festivals of light. Tulsi plant is again associated with pious sentiments.

Lines: The spaces between two lines in the images of Alpana and also the folio of Bal-Gopal-Stuti are filled with repetitive smaller lines which gives a sense of rhythm and harmony to the artwork. These lines are simple formation of geometrical shapes, which are inspired and influenced by nature, beliefs, culture of the society or community. Three lines symbolise the trident and also the grass or crops, depending on the thickness of the lines and also their directions like vertical placement. The curved lines denote the waves, winds and create a sense of rhythm. The zig-zag lines indicate the mountains. The chequered lines also called jali design is frequently used to fill in the spaces.

Dots: All the elements in folk art depict Indian symbols and images. Big dots generally are used to show the sun, moon or celestial bodies, smaller dots show the beginning or starting of life like seeds. The lines are formed by dots, which is the symbol of progress and growth. In the above images of Bal-Gopal-Stuti the textile designs created has lot of dots, with slight variations, and similarly with the Alpana designs can be also seen because of dots in them.

Triangles: In the images of Alpana we can see the triangles used around circles, and also in making human figures. Similarly, in the painted folio of fourteenth century there are patterns created by using triangles. Triangles denote the past, present and future, and also *the Tridev* i.e., Brahma, Vishnu, and Mahesh, the three gods. They also signify males and females. The mountains, pointed trees, *Teeno Lok* is also denoted by Triangles.

Squares and Rectangles: These specify the land, sacred area or farm inhabited by the community. These spaces are marked by their personal beliefs and culture and are called as *chauk*. Meaning in folk arts all these symbols are used by the folk artist of various regions with slight variations and their imaginations influenced by the society.

Circles: These are drawn as holy pedestals during festivals like *Laxmi pooja* and *Saraswati pooja* with footprints of goddess in the centre. Circular or spiral lines represent continuity, universe or space. In the Alpana samples there are many patterns created by circles, likewise there are designs created by circles in Bal-Gopal-Stuti, like the trees, jewelleries etc.

CONCLUSION

Motifs and patterns similar to Alpana folk art of Bengal are clearly in the Bal-Gopal-Stuti paintings. While painting Bal-Gopal-Stuti, the artists have employed the folk elements keeping in mind the culture and traditions prevalent during those times and must have been influenced by the other art forms like sculpture, dance, architecture and drama to have inculcated the folk elements. Folk arts play an instrumental role in reinforcing national integrity, crystallizing social solidarity, fortifying communal harmony, intensifying value-system and promoting the elements of humanism among the people of the country. These arts have been with mankind since nomadic times. The artist has taken inspiration from nature and the world around him. The motifs, elements prevalent at those times are still used in contemporary folk arts.

This art is not based on bookish knowledge but has passed from generation to generation and has its deep roots in age-old experiences. Exposure to these folk arts has inspired many contemporary and modern artists. Although considered childlike and simplistic, folk paintings are not just decorative, but carry deep cultural and symbolic meanings, aesthetic delight and the essence of spirituality which inspires imagination and are essentials of a fully evolved persona. These folk arts as well as miniature paintings need to be conserved and their significance needs to be acknowledged for future generations to comprehend the cultural roots and rich heritage of our country.

REFERENCES

[1] Harsha, V. and Dahejia, J.B. (2019). *Balgopalstuti Celebrating The Child Krishna.* New Delhi: DK Print World.

[2] Kramrisch, S. (1986). Painted Delight, Indian Painting from Philadelfia collection. Philadelfia Musuem of art.

[3] Pandey, A.M. (2020). Some Repesentative Folk Art of India. *International Journal of Research - Granthaalayah* 8(3), 348–356.

Forgery over Genuine Signature

Ritu Bharti[1], **Kavita Sharma**[2] **and H.S. Tuteja**[3]

ABSTRACT

Many times a disputed document bears genuine signature but the actual person whose signature is on the document denies that he has never signed such document. Forgery of this kind is prevalent in numbers of cases. The forger anyhow obtains a genuine signature either by request being known to the forger or by chance gets some blank paper which bears his genuine signature. After that the signed document is utilized by him to create forgery. This research paper focuses on the forgery by a fraud in which forged document has been prepared by the forger over genuine signature.

Keywords: Disputed Document, Forger, Forged Document, Fraud.

INTRODUCTION

Any document with suspected origin, contents and manner through which it is produced fall under Questioned document (Osborn, 1929). These types of documents are not genuine because it is made fraudulently for gaining some benefits. The forgery is as ancient as writing and in a number of ways a forged document can be produced and updated with time and era (Harrison, 2003). It is necessary to consider the characteristic features of genuine signature to distinguish a forged writing/signature from genuine one. Documents feature prominently the financial, legal, business, social and personal affairs. Also these documents are involved in fraud, forgery, counterfeiting, impersonation, blackmail, threat or murder, etc. (Bisesi, 2006).

Many different techniques are adopted to produce a forged document. The easiest process is the process of tracing where the forger has the model of genuine signatures. The forger does not know that it is the most easy to detect. In such cases the forger does not possess the necessary skills to adopt line quality, speed, and manner of formation of letters and to the spacing and connecting stokes in genuine signatures. The simulation process of forging signatures is the most difficult process and requires writing skills of superior class/grade.

---

[1] Research Scholar, Shri Vaishnav Institute of Forensic Science, Shri Vaishnav Vidyapeeth Vishwavidyalaya, Indore, Madhya Pradesh, India. E-mail: ritub183@gmail.com

[2] Professor, Shri Vaishnav Institute of Forensic Science, Shri Vaishnav Vidyapeeth Vishwavidyalaya, Indore, Madhya Pradesh, India. E-mail: kavitasharma5872@gmail.com

[3] Retired G.E.Q.D. Expert, Shimla.

Another easy method to prepare the forged document is forgery over genuine signature. It will depend upon the ingenuity to prepare a genuine signature on a plain paper/stamp paper with some excuse. Forgery can also be done through transplant process from photocopied method and by removing the original writing on some document through chemical erasure & then prepare forged document. Many other types of forgery and purpose related with it are described below:

1. Sometimes a signature is obtained from old relative for preparing some application to be submitted in government department/corporation for sanction to do repair/addition in building plan etc.
2. Sometimes a signed plain paper/stamp paper is given to close relative to give to advocate for submission of an application, affidavit etc.
3. In case of requirement of loan.
4. To get signature of a recruited employee on the day offering him job in office/ factory.
5. Staff member may get a genuine signature of the proprietor/Sr. officer, keep it between various letters for signatures at lower location of the page.
6. To make use of genuine signature signed with much space left in between the text written typed above.
7. To make use of blank space left between.

The present paper focusses on forgery over genuine signature where two examples are discussed one is Stamp Paper and the related bank cheque and second is a Receipt.

PURPOSE OF EXAMINATION

1. Whether the disputed cheque was written for Rupees 4, 20,000/- or some alterations have been made in the figures of amount.
2. Whether the writings over the cheque were written subsequently on the blank cheque bearing signature.
3. Whether the signature of Mahant Bhagwan Das was obtained previously and then the text of guarantor has been typed subsequently.
4. Whether the signature and the text of the cheque have been written with the same Ink/Pen.
5. Whether the receipt obtained is authentic or any alterations were made in its content.

PROCESS OF EXAMINATION

All the disputed Documents i.e., cheque for Rs. 4, 20,000/- as well as signature and text on the stamp paper and receipt have been carefully and thoroughly examined with the help of scientific equipment like magnifying lenses of various powers, measuring scale, microscope as well as through computer and scanning process.

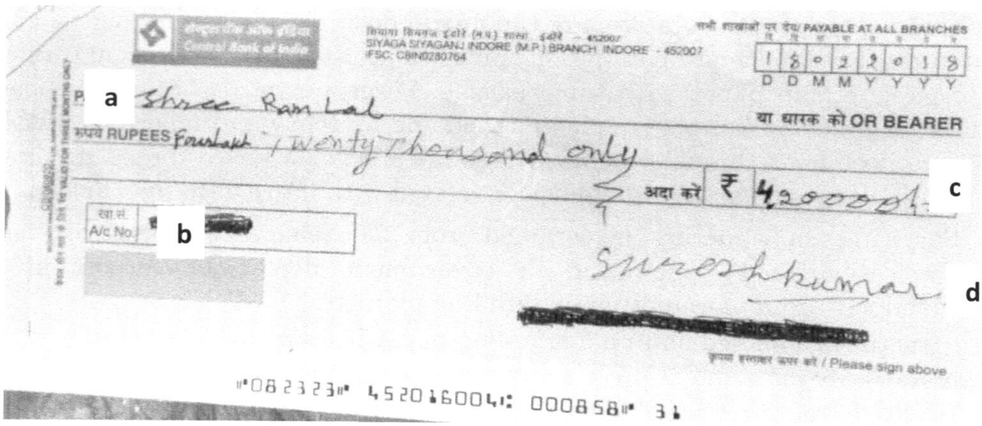

Fig. 1: Questioned Cheque, a: Dark blue pen, b: Squeezed letter, c: Subsequently added '4', d: Signature (in light colored pen same as "figure Twenty thousand only")

OPINION

1. The body writings on the disputed/questioned cheque (Figure 1) have been written with dark ink where as the tint and luster of the signature is light sky blue color.
2. Careful examination shows that the figure of 'four (4)' of 4,20,000/- has been squeezed on the left side of the amount indicating that it has been added subsequently. It appears that the cheque was originally written for only Rs. 20,000/- and has been altered to 4,20,000/-.

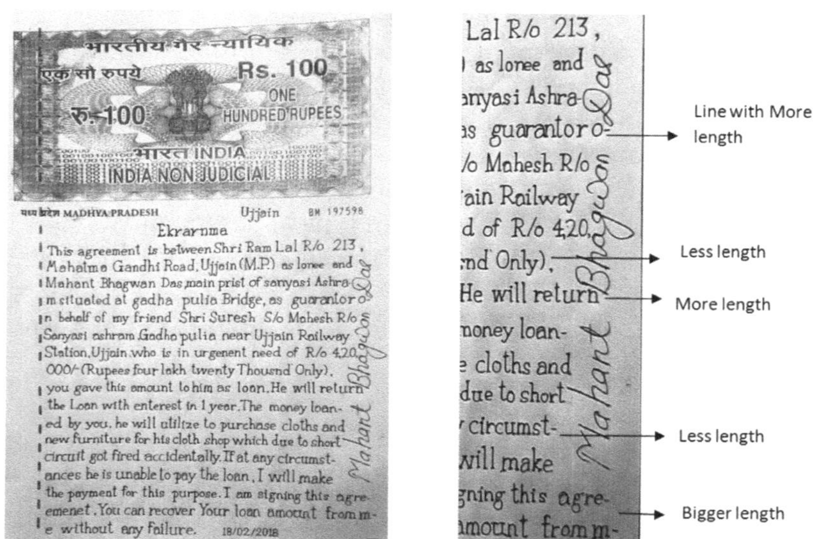

(a) (b)

Fig. 2: (a) Questioned Stamp Paper, (b) Forged Signature on Stamp Paper

3. The English text over the stamp paper shows exact and equal margin on the left side and if there is no obstruction then the right margins of the typed text will also have almost same margin except 2–3 letters will be extended for completing the word.

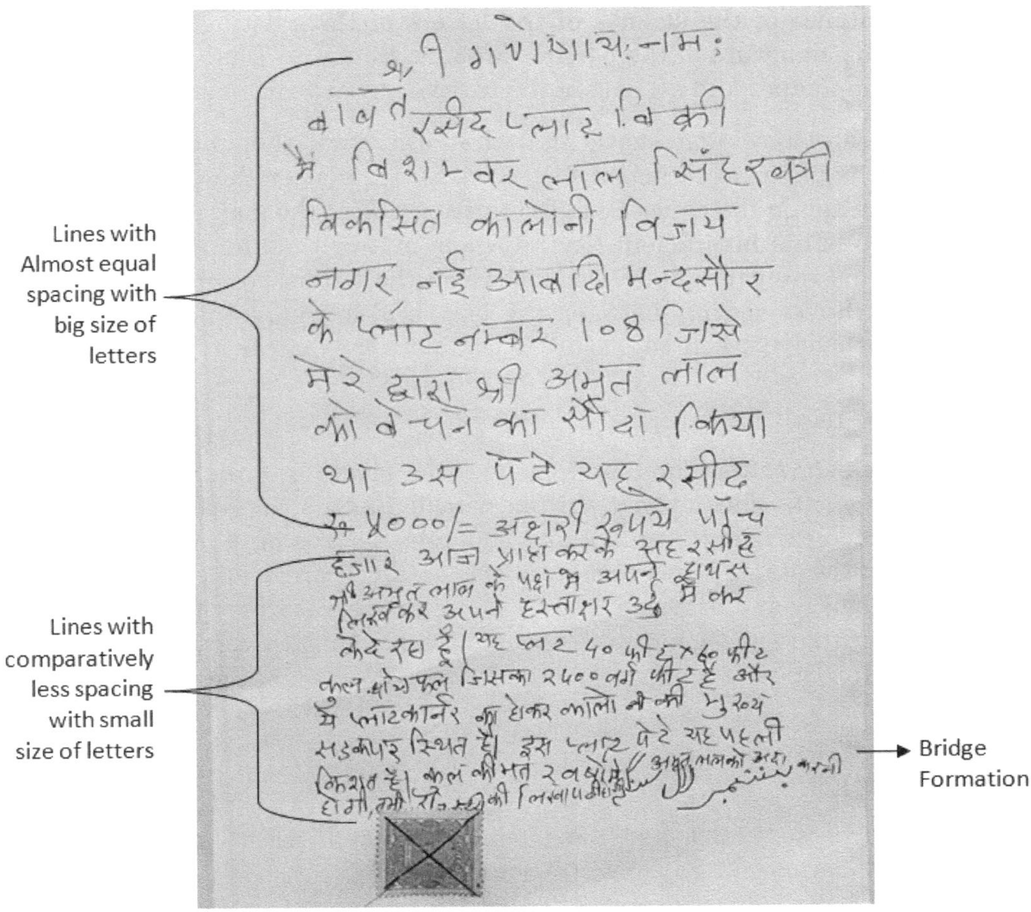

Fig. 3: Questioned Receipt

4. In receipt it shows that in the starting 10 lines there is equal spacing with big size of letters and in rest of the 9 lines, spacing is less with small size of letters.

DISCUSSION

The text which has been typed on the stamp paper shows uneven length of typed lines. Wherever the letter of the signature is of bigger size, the length of the typed lines becomes small but when the size of the letters for example a, c, i, m, etc is there, the length of typed lines will be more. Such condition of the typed text has been observed on the stamp paper. As soon as the text reaches the 1st letter of the signature, all typed lines below the signature show bigger length and are almost of the same size and reached to the right leaving little margin. These conditions have been observed in the typed text. The reason of uneven length of the typed text is due to existence of the strokes of the letters of the existing signature. This proves that the signature of the person "Mahant Bhagwan Das" was obtained on the blank stamp paper and all part of the text has been written subsequently.

In the case of receipt it shows that the forger first thought that the space provided on the paper is sufficient for the complete matter therefore, he started writing the matter on the paper and after writing half of the matter when he found that the space left is insufficient for the whole matter then he tried to cover his whole matter on the paper as a result of which the letters became comparatively very small with less spacing between the lines and in the end he made a bridge above the previously present signature to complete the matter.

CONCLUSION

Above findings show that the blank cheque was given for only Rs. 20,000/- and the signature on the cheque was written with light blue colored pen and rest of the writing (Body writing), figures and squeezed figures of '4' was also written afterwards. In the stamp paper signature was obtained previously and matter was added subsequently and in receipt the Urdu signature was taken fraudulently and matter was written subsequently. Genuinely such condition occurs when writing the matter after having signature, constitute a form of Forgery over Genuine Signature.

REFERENCES

[1] Bisesi, S.M., Kelly, S.J. and Lindblom, S.B. (2006). *Scientific examination of questioned documents*. Taylor & Francis Group, LLC.

[2] Harrison, R.W. (2003). *Suspected documents*. Universal Law Publishing Co. Pvt. Ltd.

[3] Osborn, S.A. (1929). *Questioned documents*. Boyd Printing Company.

Fragmentation and Liberty in Doris Lessing's *The Golden Notebook*

Pamila Neema[1] **and Anurag Joshi**[2]

ABSTRACT

In the wake of transformation and change in the post war era, the status of women is thought to have changed at every level of human survival. Women are thought to be the instruments to bring more stability to family, community and subsequently to the society. The isolated existence and idea of freedom is a new concept in womens' lives in the post war era. The present study explores Doris Lessing's most discussed heroine Anna Wulf in The Golden Notebook (1962) from the perspective of fragmented identity of a woman in the post war era.

Keywords: Fragmented Identity, Freedom, Isolated Existence, Post War Era.

INTRODUCTION

The post-war period had brought various social, economic and political changes in the society. Every war has both positive as well as negative outcomes. The level of destruction, distress and devastation to the society varies and therefore the reaction of people towards social problems also varies. It is difficult to live war times and endure post war expectations of new social standards. In the wake of transformation and change in the post war era, the status of women also changes. Women are thought to be the instruments to bring more stability to family, community and subsequently to the society.

The struggles of war, industrial revolution and scientific advents changed social, economic, and political scenes of the countries. Social institutions started to redefine themselves. Social classes, family's ethnicity and other institution of the society underwent many changes. War and violence brought institutional changes, which resulted in women getting lesser opportunities to grow in their social status within their communities. As women were associated with procreation and household responsibilities, it was the responsibility of man to handle and contribute to the nation building. When Doris Lessing started writing in 1950, the world of women was confined to the responsibilities of the household,

---

[1] Research Scholar, Shri Vaishnav Institute of Social Sciences, Humanities and Arts, Shri Vaishnav Vidyapeeth Vishwavidyalaya, Indore, Madhya Pradesh, India. E-mail: pamilaneema2972@gmail.com

[2] Associate Professor, Shri Vaishnav Institute of Social Sciences, Humanities and Arts, Shri Vaishnav Vidyapeeth Vishwavidyalaya, Indore, Madhya Pradesh, India. E-mail: anu_rag1978@rediffmail.com

nurturing the kids, and maintaining the family relationships and the most challenging was adapting to the modern thoughts.

A female writer, Anna Wulf is the protagonist of *the Golden Notebook*. Anna is a single mother. The man she loves abandons her for marriage. She finds that she is free but feels herself loveless .She used four notebooks to record her life and - a golden notebook to amalgamate the four notebooks to gather her own self. One can see it from the notebooks women's inner world and women's complex life in every aspect. The novel shows the protagonist's true life portrayed in her notebooks which are in parts as her own life. Anna's recording in the black notebook is collection of her African experiences; red notebook records liberal women's intuitive political views; yellow is a novel written by Anna; and the blue is a personal diary.

Anna believes that she could help herself from the chaos of reality and her four notebooks preserve her life's integrity. But she found herself getting splitted into various roles. Lessing gives the idea of love as a matter of urgency in the novel. Anna has various roles: mother, writer, and an independent woman. These roles and emotional obstacles cannot be coordinated by Anna when she is alone though she vents herself out in the notebooks, but outer chaos is much disturbing to restore her inner self.

Anna almost gets insane because she was politically disappointed by communist party and her lover abandoned her. She has been looking for a psychiatrist's assistance however, it was not helping, so she immersed herself in her private diary. With the assistance, Anna recognized her writing obstacles and actively sought new selves from Saul, a writer. Anna finally coordinated various roles and went back to normal life, childcare, job finding, overcoming obstacles to writing, and finishing the news. In every respect, Lessing reflects Anna's life in notebooks. Anna wishes to integrate her fragmented experiences, but undergoes crisis in her own identity and deterioration aggravates. Anna's fragmented state of being integrates with the help of Saul Green and the chaotic status is recovered. The recovery of Anna and Saul proves that man and woman relationship needs sacrifices and cooperation.

The theme of *The Golden Notebook* revolves around the men and women relationship, political disillusionment and disappointment, the pursuit of freedom and the self. The fight against males and their ways to treat women in the novel gave great popularity to the book as a feminist epic amidst patriarchal society. The most remarkable thing about the novel is that it describes women's real-life experiences and dilemma in seeking freedom and equality.

According to Lessing, *the Golden Notebook* was not a clarification call for the liberation of women, but still readers feel that she was concerned about women's destiny and reflected all the sympathy for women in her novel.

CONCLUSION

The Golden Notebook describes women's lives, mental conditions and behaviour vividly. The character of Anna Wulf explores liberal women's lives that depicts their painful experiences and dilemma on several outcomes. While striving for freedom and equality and managing work, marriage and kids, women are thought to restore mental balance in an idealised way. Lessing asserts that women who want freedom and equality in a patriarchal society need to come out of their traditional mind-sets. They need to be aware of their strengths and not undermine themselves as inferior to men. While doing their traditional roles they need to pursue their quest for freedom. Pursuit of freedom does not mean that they have to run away of their responsibilities towards the family and the society rather they have to realise that they have the capability of playing various roles simultaneously which elevates them to a status that is higher to men even in patriarchal society.

REFERENCES

[1] Feigel, L. (2018). *Free Woman: Life, Liberation, and Doris Lessing.* Bloomsbury Publishing USA.

[2] Joannou, M. (2000). *Contemporary Women's Writing: From The Golden Notebook to The Color Purple.* Manchester UP.

[3] Lessing, D. (2012). *The Golden Notebook.* Harper Collins UK.

Impact of Petroleum Soil Pollution on Fungal Community

Tufan Benal[1] **and Smriti Chitnis**[2]

ABSTRACT

Soil pollution has become a serious problem. Increasing population has led to establishment of many industries to meet human needs. Many industries are related to petroleum, crude oil, or hydrocarbons. Due to the leakage of petroleum products in the surrounding, soil is polluted. There is considerable variation in cartilage mycoflora in petro polluted sites and control sites. Members of some mycoflora such as Penicillium spp., Rhizopus spp., Aspergillus spp., Cladosporium spp. removes hydro-carbon from the soil, thus improving soil fertility gradually. These fungal cultures can be employed to remediate petro-polluted soil and therefore, can be a good alternative as soil reformers.

Keywords: *Aspergillus sp.*, Crude oil, Mycoflora, *Penicillium sp.*, Petroleum Hydrocarbon.

INTRODUCTION

At present, soil pollution is becoming a serious problem because due to the increasing population, human beings have created many industries to meet human needs, many of which are related to petroleum and petroleum products, crude oil, or hydrocarbons from these industries. Due to the leakage into the surrounding soil it gets affected. Petroleum soil pollution is having a detrimental effect on the fungal community and the plant community, because the hydrocarbons reduces the fertility of the soil (Das and Mukherjee, 2007). The changing chemical nature of the soil also affects the fungal community. The present paper studies the polluted areas, the growth of fungal community, their diversity, and the changes in soil. Previous researches have shown that the agricultural land and crops in such areas are getting affected due to more number of petro filing stations in such areas. Many changes have also been observed in the variety of fungi found, because of the effect of hydrocarbons on the chemical nature of the soil.

---

[1]Research Scholar, Department of Botany, Shri Umiya Girls College Rau, Indore, Madhya Pradesh, India. E-mail: deepbenal10@gmail.com

[2]Professor, Department of Botany, Mata Jijabai Govt. Girls College, Indore, Madhya Pradesh, India. E-mail: smritichitnis1@gmail.com

At present, India has many petroleum depots, refinery, and petrol pumps, due to which the soil is getting polluted with hydrocarbons. This impact is upto 2–5 Kms of the soil around the petroleum depot and petroleum refinery and therefore the mycoflora is also getting affected. There is considerable variation in the mycoflora in petro polluted areas when compared with control areas. Members of some mycoflora such as *Penicillium* sp., *Rhizopus* sp., *Aspergillus* sp., *Cladosporium* sp. (Mancera-Lopez & Rodriguez-Casasola, *et al.*, 2007). Hydrocarbon when removed from the soil can improve the soil fertility gradually. In the future, petro-polluted soil can be improved through these fungal cultures; so, it can be a good alternative to soil reformers. In general, this study has shown that more fungal diversities are found in petro-polluted soils, which confirms that many fungal species can digest hydrocarbons, that is why fungal diversities are found more in petro-polluted soils. Excess of fungal diversity can be considered as a good indicator for ecology in petro-polluted soil, but many fungal species are fatal to humans and plants as many types of fungi act to spread diseases in humans and plants.

Research Site

Some farmers of Mangalia village were complaining that the land around their village Mangalia depot has seen a lot of change in the production of crops and from earlier studies, it was found that the soil is polluted with petroleum. Many changes were also observed in the diversity of the mycoflora in the soil, so the present study was taken to study the changes in the biodiversity of the mycoflora and the changes in the soil.

SAMPLE COLLECTION

The sample was collected from:

- Polluted sites chosen were Mangaliya Depot Indore and Petro-filing Station. The polluted site was located around 15 Km away from Indore City (at Mangliya Gaon)
- The control site was the area around 2 km from the polluted site.

Mangalia Depot Indore: Bharat Petroleum Corporation Limited (BPCL). This village is in Mangalia, Tehsil Saver, and District Indore (MP). This depot is located on National Highway-3, 15 km from Indore city, this terminal was established in March 1992. It is spread over about 13.5 acres, along with Indian Oil Corporation Limited (IOCL) and Hindustan Petroleum Corporation Ltd. (HPCL) is also located. The total storage capacity of this terminal is 22,196 kiloliters. There are a totals of 16 tanks in this terminal, in which (10 underground and 6 above ground), oil is stored in them and from this, oil is supplied to cities and states. MS (Motor Spirit), HSD (High-Speed Diesel), SKO (Superior

Kerosene Oil), Hexane, Furnace Oil, and Ethanol are supplied from here. Oil is distributed through tankers. Bharat Petroleum Corporation Limited is a public sector government company that sells refined crude oil. BPCL is included in India's second-largest Navratna Company (Environmentclearan.nic.in/PREFEASIBILITY, 2017).

MATERIALS AND METHOD

Soil Sampling and Analysis

Soil samples were selected from designated sites like (Petro Polluted Site and Control Site) Soil testing was done in validated laboratories such as Indian Soil Testing Laboratory, Bhopal, and Agricultural Soil Testing Laboratory (Government Agricultural College, Indore). Soil testing was conducted on the following parameters such as pH, salinity, organic matter, nitrogen, phosphorus, potash, and carbon, etc. (soilhealth5.gov.in.googlescholar). Based on the results of these tests, changes in the diversity of the fungal community in petroleum polluted soil and controlled soils were studied.

Table 1: Results of Soil Analysis

S. No.	*sample*	*(pH) (1:2)*	*(EC) (1:2)*	*Carbon Concentration*	*Available Nitrogen*	*Available Phosphorus*	*Available Potash*
			DSM-1	%	Kg/ha	Kg/ha	Kg/ha
01.	Control site (Gram Field)	8.06	1.24	0.60	315	16.8	640
02.	Petro Polluted site (Depot Soil)	7.7	0.62	0.90	284	13.6	560

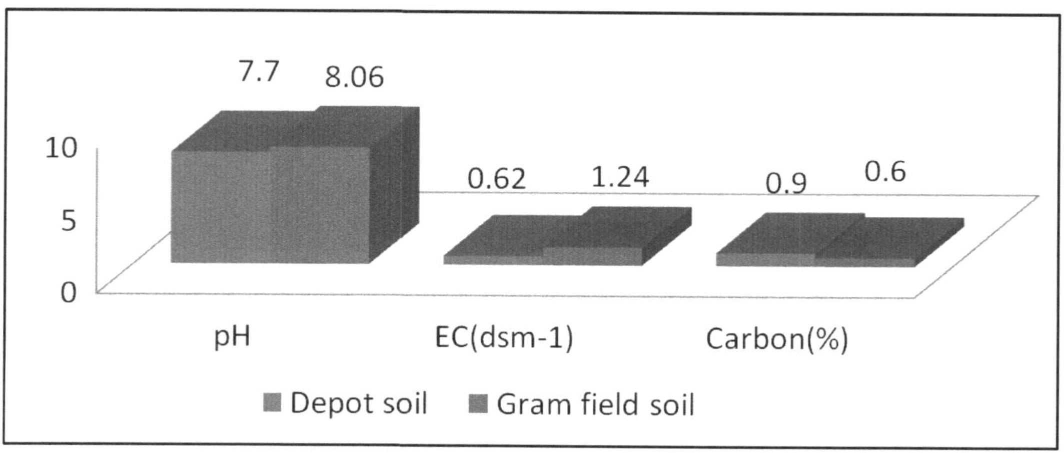

Fig. 1: Analysis of Soil pH, EC and Carbon

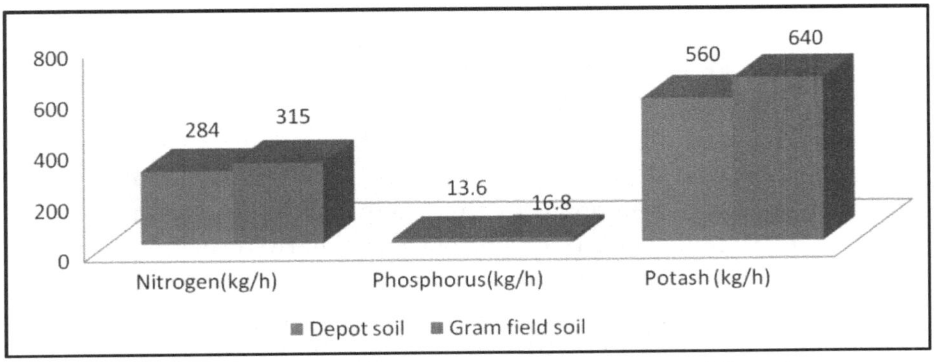

Fig. 2: Analysis of Soil Nitrogen, Phosphorus and Potash

Identification of the Soil Fungi

Potato dextrose agar media (PDA) was used to isolate the fungi present in the soil by serial dilution technique. In this culture media, the fungi grow easily and colonies of many fungi are seen on the Petri plate. Members were identified based on the color of the colony, size of the colony, the arrangement of mycelium and conidia. Standard literature and Google scholar's help was taken to identify fungi species. Lacto phenol cotton blue was used for microscopic slide staining.

RESULTS AND DISCUSSION

It was found that fungal diversity is seen more on the petrol-polluted site, whereas normal fungal diversity is less frequent at normal sites. Some species of fungi are also found that consume hydrocarbons. While some species of fungi also show higher growth on hydrocarbons. 10 different types of fungal species were isolated from Depot soil and Gram field soil. They were like *Aspergillus* sp., *Penicillium* sp., *Fusarium* sp., *Trichoderma* sp., *Alternaria* sp., *Curvularia* sp., *Mucor* sp., *Rhizopus* sp., *Morssonina* sp., *Helminthosporium* sp. Similar findings has been reported by Cerniglia (1997) and Chaillan *et al.* (2004). The study reinforces the earlier findings that due to the current climate change, types of changes are being seen and all of them are due to increasing industrialization which results in different types of pollution.

Through this study, it has been found that fungal species are found much within the radius of 2 km from the petrol-polluted site. Plant growth is slow due to oil pollution, and the growth of fungi present in the soil is also slow but not in all the fungi as some fungi can break the chain of hydrocarbons (Reiche and Lemos, 2006). Relative to the diversity of fungi in hydrocarbon polluted soils and similar results have been found in research in which fungi species such as *Penicillium* sp., *Aspergillus* sp., *Fusarium* sp., and *Mucor sp.*, were found in abundance in the polluted soil (Obire *et al.*, 2009).

CONCLUSION

This research was carried out in the soil of two major sites, one of which was the petro-polluted site Mangaliya Depot Indore and the other sites was controlled sites (gram fields). The pH of these two sites, the amount of the organic matter, nitrogen, phosphorus, and potash in the soil was analysed. A comparative study of the presence of fungi was done. In this comparative study, it was found that the growth of the fungal community was observed more in the petroleum polluted soils whereas, the growth of the fungal community at the controlled sites was seen within the normal range. Inferior growth of fungi like *Aspergillus* sp., *Penicillium* sp., *Fusarium* sp., *Trichoderma* sp., *Alternaria* sp., *Curvularia* sp., *Mucor* sp., *Rhizopus* sp., *Morssonina* sp., *Helminthosporium* sp, *Cladosporium* sp., and *Chaetomium* sp., was observed. But some of these fungal species can digest hydrocarbons such as the *Aspergillus* sp., *Penicillium* sp., *Fusarium* sp. These species increase the fertility of the soil by reducing the number of hydrocarbons in the soil and can help in the future, to improve the quality of petro-polluted soil.

REFERENCES

[1] Chaillan, F., Fleche, A.L., Bury, E., Phantavong, Y., Crimount, P., Saliot, A. and Oudot, J. (2004). Identification and biodegradation potential of tropical aerobic hydro-carbon-degrading microorganism, *Research in Microbiology,* 155, 587–595.

[2] Das, K. and Mukherjee, A.K. (2007). Crude petroleum-oil biodegradation efficiency of *Bacillus subtilis and Pseudomonas aeruginosa* strains isolated from a petroleum-oil contaminated soil from North-East India, *Int. Res. J. Biological Sci.* 3(11) 28–31.

[3] Mancera-Lopez, M.E., Rodriguez-Casasola, M.T., Rios-Leal, E., Esparza-Garcia, F., Chavez-Gomez, B., Rodriguez-Vazquez, R. and Barrera-Cortesa, J. (2007). Fungi and Bacteria isolated from Two Highly polluted Soils for Hydrocarbon Degradation, *Acta Chimica Sloenica,* 54(1), 201–209.

[4] M/s Bharat Petroleum Corporation Limited (2017). Environment Clearance Manglia Depot is located at Village Mangalia. Tehsil-Sanver, Agra Bombay Road, P.O.2017.from http://www.environmentclearance.nic.in

[5] Oboh, O.B., Ilory, O.M., Akinyemi, O.J. & Adebusoye, A.S. (2006). Hydrocarbon degrading Potentials of bacteria isolated from a Nigerian Bitumen (Tarsland) deposit *nature and Science*, 4(3), 51–57.

[6] Obire, O. and Anyanwu, E.C. (2009). Impact of various concentrations of crude oil on fungal populations of soil *International Journal Environmental Science and Technology.,* 6 (2), 211–218.

[7] Reiche, A.P. and Lemos, J.L.S. (2006). Estudo do potencial de degradação de petróleo de linhagens de fungos isoladas de solo nordestino. In: XIV Jornada de Iniciação Científica, Centrode Tecnologia Mineral – CETEM/MCT. 55.

[8] Soil Testing Laboratory Report – Soil Health Card/ Soil Testing Laboratory Report. Sr. No. Lab Id. State. District. Lab Name. 1. 2. 3. 4 ... Indore. Soil Testing Lab (Agriculture College) Indore. 929. 2253.from http://www.soilhealth5.gov.in

Instantiation of Anglo-Irish Relations in William Trevor's *Big House Novel—The Story of Lucy Gault*

Akanksha Shekhawat[1] and Roopa Shinde[2]

ABSTRACT

The word "big house" applies back to Ireland's castles, villas, mansion and country houses that are regarded as the Anglo-Irish class itself. In Ireland, Anglo-Irish was the dominant class because of the Protestant Supremacy. This genre has traditionally been used to represent the struggles of the Anglo-Irish protestant family landowners. Four prominent characteristic features of Big house literature according to Vera Kreilkamp are - Big House novels present the decaying house as a symbol of the declining landed class; they include problems related to the lineage and succession of the Big House family; they portray a major character as an irresponsible, exiled landlord; and they introduce a character from outside the demesne wall who eventually gains control within (Kreilkamp 24). The paper aims to analyse William Trevor's portrayal of Anglo-Irish relations in his Big house novel The story of Lucy Gault (2002) which is set amid the tension between England and Ireland in the early 20th century.

Keywords: Anglo-Irish, Big House Tradition, Catholic, Failure, Protestan.

INTRODUCTION

The contemporary novels are primarily attributed to the Big House literary tradition. The Big houses of Ireland are the mansions, country houses built by the aristocrats in the mid 17th century and are regarded as the Anglo-Irish class itself. The British Government awarded acres of land to the members of the Anglo-Irish community after which they soon became the economic and political leaders of Ireland. The big houses established by protestant aristocrats reflected their dominance and were intended to create respect and honour among the middle- and lower-class citizens. The houses became the symbolic key element of Ireland's Anglo-Irish hegemony at the end of the 16th century, and during the Irish revolution many houses were burned or destroyed.

[1] Research Scholar, Shri Vaishnav Institute of Social Sciences, Humanities and Arts, Shri Vaishnav Vidyapeeth Vishwavidyalaya, Indore, Madhya Pradesh, India. E-mail: aashi.2493@gmail.com

[2] Associate Professor, Shri Vaishnav Institute of Social Sciences, Humanities and Arts, Shri Vaishnav Vidyapeeth Vishwavidyalaya, Indore, Madhya Pradesh, India. E-mail: roopashonde@svvv.edu.in

The story of Lucy Gault emphasizes the symbolic representations of failure and sufferings. Trevor's fiction includes certain elements of this genre such as abandoned big houses, rural landscapes, the disappearance of the protestant landlords and the emerging middle-class Catholicism. Though his novels undermine the attributes of the Big House tradition, he re-establishes this attribute with innovative narrative methods. The novel depicts the lives of a number of citizens implicated in 1921 attack by attackers on the Gault family. The troubled relationships between natives of the country in the year 1920 are depicted in this novel. Anglo-Irish ascendancy has been favoured in the past because of their protestant religion and connection with England. The novel is set in Cork in the big house named Lahardane. This had been the home of the Gault family since centuries.

The novel starts when the protagonist Lucy was just eight years old. Lahardane was attacked one night in June 1921 and Captain Everard Gault, the owner of the house, mistakenly shot one of the attackers. He and his wife Heloise, apprehending the war on their family, plan to flee the country with their daughter Lucy. But she did not want to leave her house therefore, she runs away to express her disagreement. Regretfully, after her escape Lucy ends up hurting her leg and is compelled to endure in an isolated area. Her parents are unable to locate her. They found her clothes on seashore and presumed that Lucy might have drowned herself. Her parents leave the country to escape the wrath of the natives. After few days Lucy was found but her parents never received the message that their daughter is still alive. The novel depicts the lives of the Gaults, Lucy and the man who was injured by Captain named, Mr. Horahan.

Despite, Gault's long family history in Cork, they became the victim of attackers and some of the neighbours consider them as rivals because of their links to England. Héloise Gault is an Englishwoman, but had been living in Ireland for years. The Gault family had migrated to Ireland from England. Heloise blames herself on the house's misfortune after the attack. The nature of the house, the possession of land even though it had dwindled, the family's army connection, would have been enough to bring that trouble in the night (Trevor 15). In the novel, the family never seems to give a picture of such family who think themselves superior to others. Captain Gault, his wife and his daughter are the only members of the entire household. They had two servants as well. When Captain shot an attacker named Horahan, he rushed to apologize to the Horahan family in spite of being aware about the extent of hatred towards him. This act shows his humility and soberness.

He knew that his apologies will not work yet he took the initiative for a peaceful solution. Many Anglo-Irish people have chosen the same way as the Gaults. They fled from their homes and Ireland to escape the threat to their lives.

The abandoned big houses are seen as symbols of their actual owners and England's dominance and injustice. Therefore, these houses were never used by others as their home, instead were viewed as undesirable objects. The exclusion of the people is carried on to the houses. Though, the shot was accidental, the Horahan family did not accept this. They don't trust his words and neither acknowledge his apologies nor the indemnification he gives them. The captain had used the pistol to scare the attackers. He didn't intend to harm anybody. The progression of years is demonstrated by incorporating the references to historical incidents, such as the eventual breakout of World War II and Anglo-Irish conflicts.

Ireland was extremely disturbed in 1921, as the pressure for national independence broke out resulting in brutality against British military and Anglo-Irish civilians, who were unexpectedly made targets for grudges that had spanned seven hundred years of injustice. Mr. Gault suffered great misery, but still he had no sense of hatred for his country and his people. His character emerges against the dominant portrayal of the family member of Ascendancy. His misfortune is seen as a result of the Irish history. Trevor had a less politically conservative outlook towards protestant tenants which a variation he experimented with the Big House literary tradition. His Anglo-Irish families do not only dwell in reduced estates, they also express sympathy to Catholics and sympathy for the Irish. Similarly, Trevor changes his own common stereotypes which fluctuate among experimental and conventional literary styles.

Lucy's endurance is shaped by a haunting enchantment that makes her resemble a secular saint. She does not surrender herself to the imprisonment and the crippling force of history, nor falls prey to the psychotic insanity that pervades Trevor's previous Big House novels. She celebrates isolation and sorrow by rejecting the proposal of Ralph. In the course of time, Lucy apparently takes her father's opinion that she doesn't have anyone to blame. Having failed to convince the Horahan's family that he had no intention of killing their son, Captain is not shocked or enraged. He sees his country as a nation with a past of healthy relationships among people. Lucy never blamed Horahan for her separation from her parents. She believes that he is a victim like her and he did what seemed to be right for others rather than himself. Several days after the meeting, Horahan loses his memory and the power of speech. He is shifted to an asylum. Lucy visits him there regularly every fortnight until his death, for the next 17 years. She used to play games with him possibly with the feeling of belongingness, or to give some sort of meaning to her solitary life (Trevor 220). Although, her motives are not specified in the novel.

Over the period of time the perception of local people changes towards Lucy. She doesn't appear to be offended by anyone because of her heritage. The hatred of others because of her foolish behaviour gave her parents much pain and frustration. As the time progresses, the mind-set towards Aristocratic families fade, as can be seen from Lucy's perspective. Trevor depicts people with diverse perceptions towards Anglo-Irish relations, although some characters come from common cultures, they are considerably distinct in their beliefs and views. Several families faced the same circumstances as the Gaults during the time of Troubles in Ireland, and were also threatened. Many fled the country and their country Big Houses that were later often demolished or abandoned.

Gaults are not surprised by the attack of arsonists on their property, since this kind of attack was usual in the region. Trevor never considered individuals for their political views. He focuses more on their experiences of the life, relationships with others and their internal world. He represents their thoughts, aspirations and fears and humanitarian fervour. Captain Gault seeking forgiveness from Horahan family, reveals the complexity of the aristocratic class. However, the Horahan's family shows the arrogance by rejecting Captain's apologies.

CONCLUSION

Trevor depicts individuals with different perspectives on Anglo-Irish relations, although few characters share common background, but their perceptions, values and attitudes are usually very different from one another. His characters are ordinary citizens, not engaged in politics or conflicts. Still these Anglo-Irish relationships have an immense impact on people's lives, mostly due to the people among them who impose the notion of common guilt. The stories are set on the backdrop of the events of the 20th century, and they contain a reference to the real events and individuals.

The self-reflective ending of the novel provides a bright contrast to pictures of misery and despair. Trevor tends to reduce the general belief of suffering and the postcolonial strain of English hegemony by turning the emphasis onto protagonist, as the title itself embodies. It is exceptional for its narrative form. Trevor demonstrates through his novel that only victims are caused by violence. Others who have nothing to do with the real crimes also become the victims such as Gault family in the novel. Even, Horahan who was praised by his people for the crime he had committed, ends up as a victim. Hence, Trevor conveys the insignificance of crime through his novel. Trevor's subjective interpretation of a nation stuck in the past is illustrated by the continuous retention of Lahardane. As Trevor transits from realism to post modernism indicates a new approach in the big house tradition. Trevor complicates the narrative, which combines historical realism with symbolism, by continuously changing the narrative focus.

REFERENCES

[1] Fitzgerald-Hoyt, Mary (2003). *William Trevor: Re-imagining Ireland*. Dublin: The Liffey Press.

[2] Foster, John Wilson (2006). *The Novel of the Big House*. The Cambridge Companion to the Irish Novel. Cambridge: Cambridge UP.

[3] Kreilkamp, Vera (1998). *The Anglo-Irish Novel and the Big House*. Syracuse: Syracuse UP.

[4] Monaco, Angelo (2016). Postcolonial Trauma in William Trevor's Anglo-Irish Big House Trilogy. *Università Di Pisa* 18, 159–172. doi:10.14277/2499-5975/Tol-18-16-11.

[5] Rauchbauer, Otto (1992). *Ancestral Voices: The Big House in Anglo-Irish Literature: A Collection of Interpretations*. Olms.

[6] Trevor, William (2002). *The Story of Lucy Gault*. Viking Press.

Integration of Edge Computing and Deep Convolutional Neural Network for Potato Plant Diseases Detection in Real-Time

Nitin Rathore[1] **and Anand Rajavat**[2]

ABSTRACT

Agriculture is the basic foundation of any nation's economy in the world. By 2050, the world needs to enhance food production to about 70% to feed an estimated overall population of over 9 billion people. Potato is consumed worldwide and its production plays an important part in agriculture. Early blight and late blight are the two main diseases that affect potato crop production severely. In order to detect these diseases in real time AlexNet, MobileNet and VGG16 three deep convolutional neural networks (CNN) were explored. All 3 CNN models were trained on cloud with a dataset of 7128 images containing healthy and diseased leaves images of potato plants. Even if training is outsourced, trained models still require a large size of RAM, so the prime goal of this work was to identify lightweight CNN that can be fitted into resource constrained devices easily. In this study lightweight deep CNN were identified and deployed at the edge node to detect diseases in real time by utilizing leaf images captured by camera devices in-place to enhance potato crop productivity and reduce economic losses. The advantage of the proposed approach is that by classifying potato leaf images in real time on-premises, there is no need to send images to the cloud, which unnecessarily increases network congestion, for possible disease detection. Results show that the identified lightweight CNN model obtained 99.87% accuracy for both train and test images and require only 22.5 MB memory. Precision, recall (sensitivity) and F1 score were also presented to visualise the model's performance.

Keywords: Classification, Cloud Computing, Deep Convolutional Neural Network, Edge Computing and Potato Diseases Detection.

---

[1]Research Scholar, Shri Vaishnav Institute of Information Technology, Shri Vaishnav Vidyapeeth Vishwavidyalaya, Indore, Madhya Pradesh, India. E-mail: nitinrathore@svvv.edu.in

[2] Director, Shri Vaishnav Institute of Information Technology, Shri Vaishnav Vidyapeeth Vishwavidyalaya, Indore, Madhya Pradesh, India. E-mail: anandrajawatds@gmail.com

INTRODUCTION

Crop production is indeed one of the major concerns for all the people because most of the food is provided through farming. In the world economy crop production plays a key role for both developing as well developed countries. In India agriculture is the prime origin of earning, on an average roughly 70% of the people rely on crop production to sustain their lives (Limbore & Khillare, 2015). Potato is consumed worldwide and its production plays an important part in agriculture. Nevertheless, in the cycle of potato crop development, many fungal and bacterial diseases pose significant obstacles (Munyaneza *et al.*, 2010). Among these serious diseases are that caused by Alternaria solani (Early Blight) and Phytophthora Infestans (Late Blight) that affect potato crop most (Henfling, 1987). Barbedo (2016) characterised the difficulties encountered in automatically detecting plant diseases, using visible images like symptoms of various diseases can be very akin and may occur at the same time.

Plant diseases were studied extensively in the scientific field, with particular emphasis on the genetic characteristics of diseases (Nutter *et al.*, 2006). The issue of plant pathogens is a global concern which is also affected due to change in climate (Coakley *et al.*, 1999). In recent years, proliferation of plant pests and diseases has grown drastically, resulting in farmers to suffer major losses. An earlier diagnosis of diseases is now a daunting method and must be handled with special care (Fuentes *et al.*, 2016). In this work integration of edge computing and deep learning approaches to detect potato plant diseases in real time was proposed. For applications that demand responses in real time, cloud computing cannot be utilized to satisfy their needs as cloud servers are located far away from end users and also cloud require high speed Internet connectivity (Bangui *et al.*, 2018). According to Zwolenski & Weatherill (2014) and Antonini *et al.* (2019) by 2025 about 45% of the world's data will be processed at the boundary of the network.

The concept of edge computing emerged to cater limitations of cloud computing by analysing, filtering and classifying data locally at the network edge. In the present work by utilising the concept of edge computing and deep CNN models, potato leaves are processed locally in order to identify possible disease type or healthy state. The prime objectives of this paper are:

1. To explore and identify deep CNN models that are lightweight (occupy less memory) for plant disease detection, so that deep CNN models can be easily fitted into edge nodes like mobile devices and raspberry pi.
2. To integrate edge computing environment and deep CNN model in a way that training of deep CNN model is outsourced to the cloud and trained model is deployed at edge node.

3. To utilise a lightweight deep CNN model to classify images of potato leaves at the edge node in real time, rather than collecting images and analysing them in the cloud which will unnecessarily increase network congestion.

REVIEW OF LITERATURE

Jadhav *et al.* (2020) presented two deep CNN models for identifying soybean diseases based on transfer learning. In their work, AlexNet and GoogleNet for soybean diseases detection achieved 98.75% and 96.25% model accuracy. Islam *et al.* (2019) demonstrated transfer learning technique using VGG16 CNN model for detecting potato diseases and achieved 99.43% accuracy. Mohanty *et al.* (2016) described transfer learning approach utilizing AlexNet and GoogleNet CNN models to classify 14 crop species into 26 classes and attained trained model accuracy of 99.35%. Rangarajan *et al.* (2018) proposed transfer learning on pre-trained VGG16 and AlexNet CNN models to distinguish tomato plant diseases and attained 97.29% and 97.49% accuracy for VGG16 and AlexNet models respectively.

In order to identify 10 common rice plant diseases (Lu *et al.*, 2017) developed a CNN model which attained 95.45% accuracy. Ferentinos (2018) presented AlexNet, GoogleNet, OverFeat and VGG16 CNN models to classify 25 different plant leaf images into 58 classes of diseased and healthy plants. In these listed CNN models VGG16 has attained best performance with an accuracy of 99.53%. Dasgupta *et al.* (2020) proposed transfer learning approach on various pre-trained CNN models like ResnetV18,V34,V50, V101, V152; Inception 4; Densenet V121, V161, V169, V201, VGG16 and VGG19 to classify potato plant diseases into 3 categories. Their results proved that VGG19 attained best accuracy. Islam *et al.* (2017) described integration approach of image processing (image segmentation) and support vector machine in order to detect potato plant diseases.

The proposed model obtained 95% accuracy in order to classify potato images. Kalvakolanu (2020) proposed transfer learning approach by exploiting Resnet34 and Resnet50 for classification of plant diseases and attained an accuracy of 99.44%. Singh & Misra (2017) introduced an image segmentation approach for detecting and classifying plant leaf diseases automatically. The above-mentioned work demonstrated implementation of machine learning, transfer learning, and Deep CNN models for the classification of plants in healthy or diseased using images of leaves. To the best of our knowledge, no one has worked out about lightweight (occupy less memory) CNN models for the classification of plant diseases, so that model can easily be fitted into resource constrained devices like edge nodes which may be a raspberry pi or mobile device. The present work identifies a lightweight CNN model and deployed it into edge nodes for real-time disease detection.

METHOD

Edge Computing Framework

In this paper edge computing framework has been utilized as depicted in Figure 1 that is made of 3 layers namely device layer, edge computing layer and cloud computing layer. Device layers consisting of camera devices like mobile phones and other digital cameras are responsible for capturing and collecting potato plant leaf images for possible disease detection. Potato leaf images collected by the device layer are fed as input into the edge computing layer. Deep learning CNN models (AlexNet, MobilNet and VGG16) trained on cloud computing layer and after that deployed at edge computing layer for detecting diseases by giving leaf images as input to mentioned deep learning models in real time.

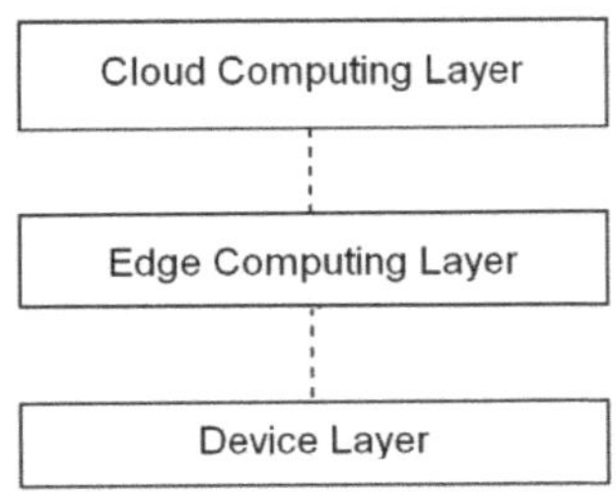

Fig. 1: Edge Computing Framework

Dataset

Potato leaves images were collected from a publicly available database (Bhattarai, 2018). In this work a total 7128 leaves of potato plant with three classes namely early blight, late blight and healthy were divided into a ratio of 80:20 for training and testing. The training dataset consists of a total of 5702 images having early blight, late blight and healthy images 1939, 1939 and 182 respectively. In the same way, the test dataset consists of 1426 images having early blight, late blight and healthy images 485, 485 and 456 respectively. Input image size in pixels is 256 × 256 × 3 where 3 represent the number of channels (RGB). The dataset summary is displayed in Table 1.

Table 1: Dataset Details

S. No.	*Disease Class*	*Training Samples*	*Test Samples*
1.	Early Blight	1939	485
2.	Late Blight	1939	485
3.	Healthy	1824	456
Total		**5702**	**1426**

Architecture of AlexNet and MobilNet CNN Model

In this work, for potato plant disease classification three deep learning CNN models were utilized. For the VGG16 model a transfer learning approach was applied and retrained only the last convolutional and output dense layer. There is no modification applied to VGG16 model. In order to accelerate the training process and boost the performance of AlexNet, MobilNet and VGG16 raw images were first pre-processed using ImageDataGenerator class of Keras.

AlexNet

Krizhevsky *et al.* (2012) proposed a deep CNN architecture to classify 1000 classes of objects in ImageNet competition later known as AlexNet. Architecture of AlexNet consists of 5 convolutional layers, 3 pooling layers, 3 dense layers including one output layer. Convolutional layer extracts activation features like edges, curves and a colour using a set of filters in the input image and the pooling layer performs down sampling on the feature map. Its goal is to gradually lessen the representation's spatial size to minimize the number of parameters and calculation within the network. The pooling layer operates independently across each feature map. The dense layer is a fully connected layer that flattens the input feature from 2D into a 1D vector, introduces non-linearity using some activation function like ReLU (Rectified Linear Unit) to categorize image-level features. In order to classify potato disease some modification have been done like changes in filter size and adding an extra dense layer to classify potato leaves into 3 categories in AlexNet architecture as represented in Table 2.

Table 2: Architecture of Modified AlexNet

Layer	*Function*	*Filter Size*	*No. of Filter*	*Stride*
Conv 1	Convolution	11 × 11 × 3	96	4
Pool 1	Max pooling	2 × 2	–	2
Conv 2	Convolution	11 × 11 × 31	256	1
Pool 2	Max pooling	2 × 2	–	2
Conv 3	Convolution	3 × 3 × 10	384	1
Conv 4	Convolution	3 × 3 × 8	384	1
Conv 5	Convolution	3 × 3 × 6	256	1
Pool 5	Max pooling	2 × 2		2
Dense-1 (4096)	Fully connected	–		–
Dense-2 (4096)	Fully connected	–		–
Dense-3 (1000)	Fully connected	–		–
Dense-4 output layer (3)	Disease classification			

MobileNet

Howard *et al.* (2017) proposed MobileNet for resource constrained devices like mobile phones. In order to build a computational efficient deep neural network, MobileNet utilises depth-wise separable convolution. Depth-wise separable convolution consists of a depth-wise convolution layer and separable convolution layer to divide the standard convolution operation into filtering features and to combine features separately. In depth-wise convolution a single filter applies to each input channel and after that to combine the output of depth-wise convolution, separable convolution applies a 1 × 1 convolution. Depth-wise separable convolution attains 8–9 times less computation in comparison to standard convolution. MobileNet architecture to classify potato leaves into 3 classes is presented in Table 3 with modified output layer.

Table 3: Architecture of Modified MobileNet

Layer	*Function*	*Filter Size*	*Stride*
Conv	Convolution	3 × 3 × 32	2
Conv dw	Convolution depth-wise	3 × 3 × 32 dw	1
Conv	Convolution	1 × 1 × 64	1
Conv dw	Convolution depth-wise	3 × 3 × 64 dw	2
Conv	Convolution	1 × 1 × 128	1
Conv dw	Convolution depth-wise	3 × 3 × 128 dw	1
Conv	Convolution	1 × 1 × 128	1
Conv dw	Convolution depth-wise	3 × 3 × 128 dw	2
Conv	Convolution	1 × 1 × 256	1
Conv dw	Convolution depth-wise	3 × 3 × 256 dw	1
Conv	Convolution	1 × 1 × 256	1
Conv dw	Convolution depth-wise	3 × 3 × 256 dw	2
Conv	Convolution	1 × 1 × 512	1
Conv dw	Convolution depth-wise	3 × 3 × 512 dw	2
Conv	Convolution	1 × 1 × 1024	1
Conv dw	Convolution depth-wise	3 × 3 × 1024 dw	1
Conv	Convolution	1 × 1 × 1024	1
Global Average Pooling	GlobalAveragePooling2D	–	–
Dense output layer(3)	Disease classification		

Experimental Setup

In this work raspberry pi 3 model B+ is used which has 1.4 GHz 64-bit quad-core processor, 1 GB LPDDR2 SDRAM and 32 GB SD card for edge node which is placed at edge computing layer. Due to intensive computing and time consuming reason AlexNet, MobileNet and VGG16 were trained/validated on a total 7128 potato leaves images at cloud computing layer having configuration of 4 core Intel Xeon CPU 2.30 GHz, 13 GB RAM and Nvidia K80 GPU with 5 GB memory. After completion of training and validation all three models are saved using a python pickle module and loaded at edge node for real time potato disease detection. Modified AlexNet and MobileNet were designed and implemented using an open source neural network API named Keras. Batch size set to 64 in Image Data Generator (for real time data augmentation and generating batches of input images) class and compiled using Nadam optimizer with learning rate 0.0001, categorical cross-entropy as loss and accuracy metrics for both the models. In order to retrain modified AlexNet, MobileNet and VGG16 on potato leaves keras.fit_generator deep learning library was utilized with number of epochs set to 15 and 10, 10 for AlexNet, MobileNet and VGG16 respectively.

RESULTS AND DISCUSSION

Model Accuracy & Loss

Training progress curves with respect to model accuracy and loss for AlexNet MobileNet and VGG16 are displayed in Figure 2(a)–(f) respectively. By observing the training progress curve it is evident that AlexNet achieves about 99.75% and 99.50% accuracy for both test and train images respectively. MobileNet achieves the same accuracy for both train and test dataset which is about 99.87% respectively. VGG16 attained about 98.84 and 99.65 accuracy for both test and train images respectively.

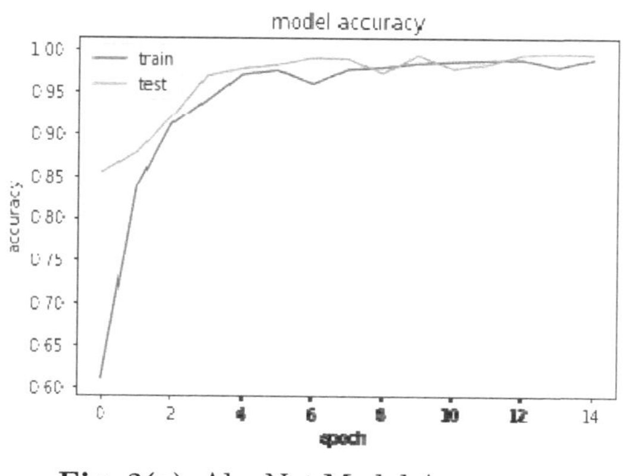

Fig. 2(a): AlexNet Model Accuracy

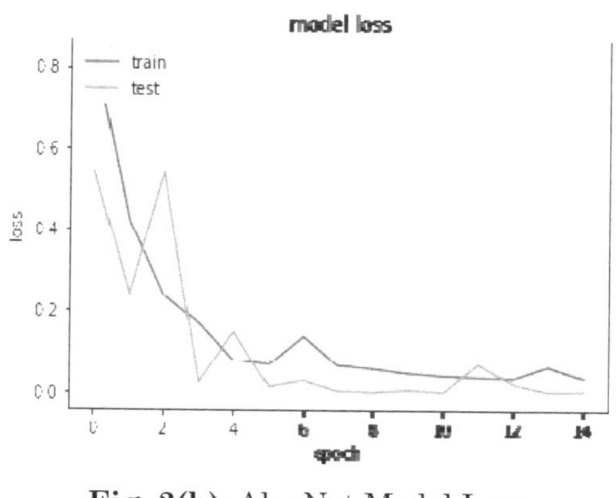

Fig. 2(b): AlexNet Model Loss

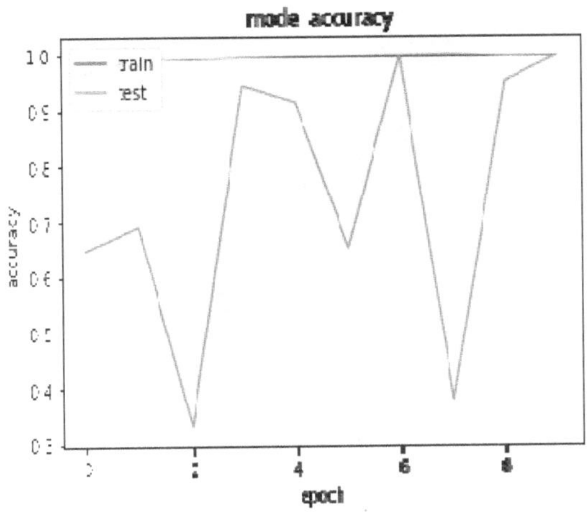

Fig. 2(c): MobileNet Model Accuracy

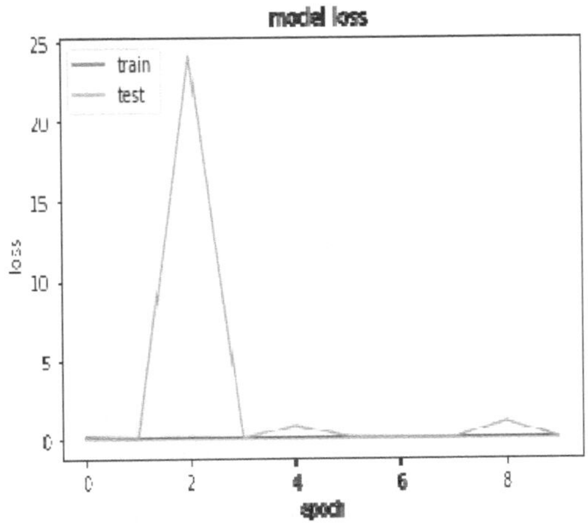

Fig. 2(d): MobileNet Model Loss

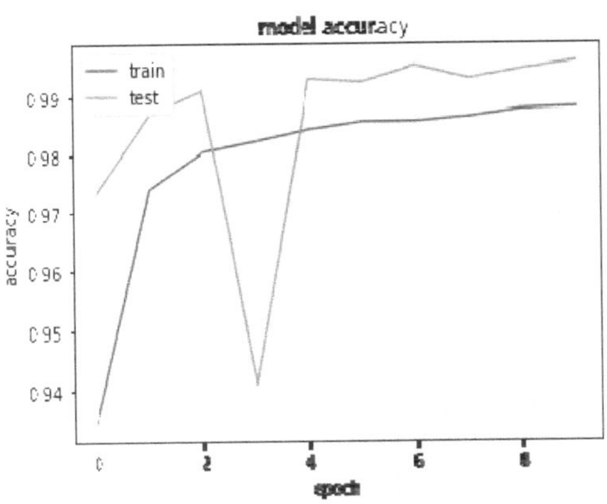

Fig. 2(e): VGG16 model accuracy

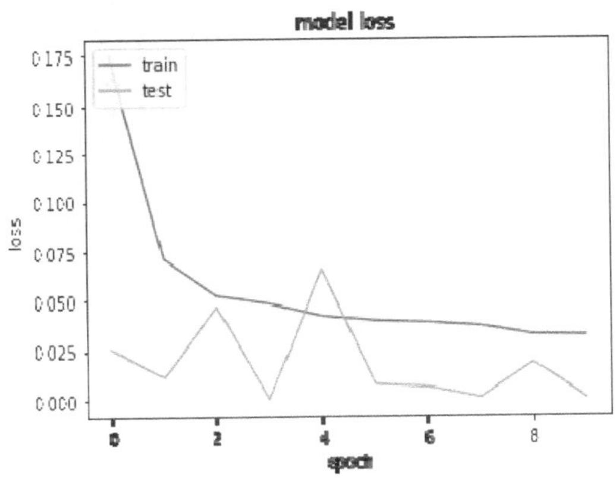

Fig. 2(f): VGG16 model loss

Fig. 2: Training Progress Curve

CONFUSION MATRIX

In Figure 3(a)–(c) confusion matrices generated for AlexNet, MobileNet and VGG16 for classification of total 1426 leaves images into 3 classes are shown respectively. In AlexNet for class 3 (healthy) all 456 images were classified correctly but for class 1 (early blight) out of 485 images 483 were classified correctly and only 2 images were misclassified. For the class 2 (late blight) out of 485 images 478 images were classified correctly but 3 images were misclassified into class 3 and 4 images were misclassified into class 1. In MobileNet for class 3 and class 1 all images were classified correctly. For the class 2 (late blight) out of 485 images 476 were classified correctly and 8 images were misclassified into class 1 and only 1 image were misclassified into class 3.

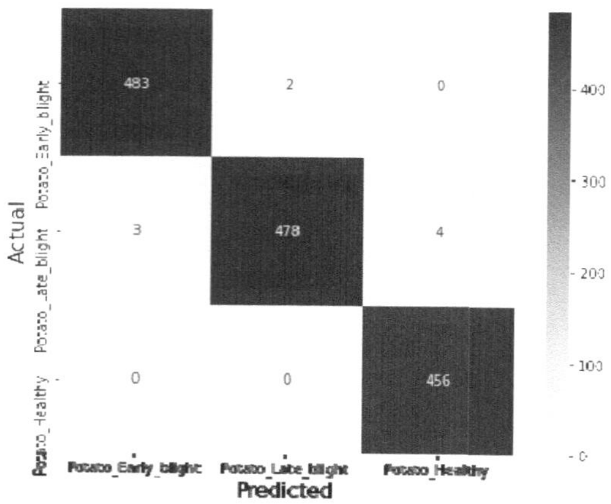

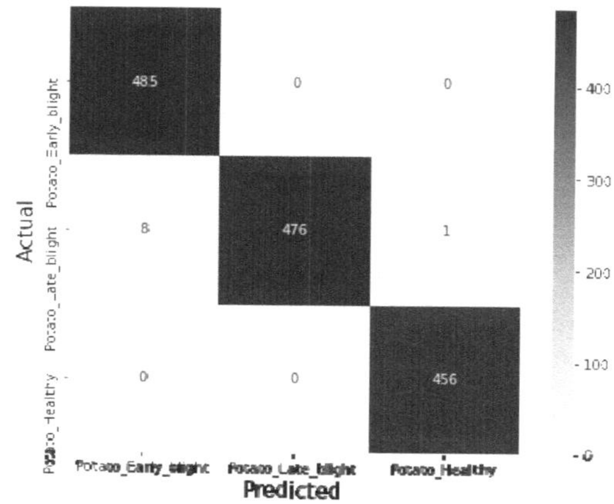

(a) AlexNet Model (b) MobileNet Model

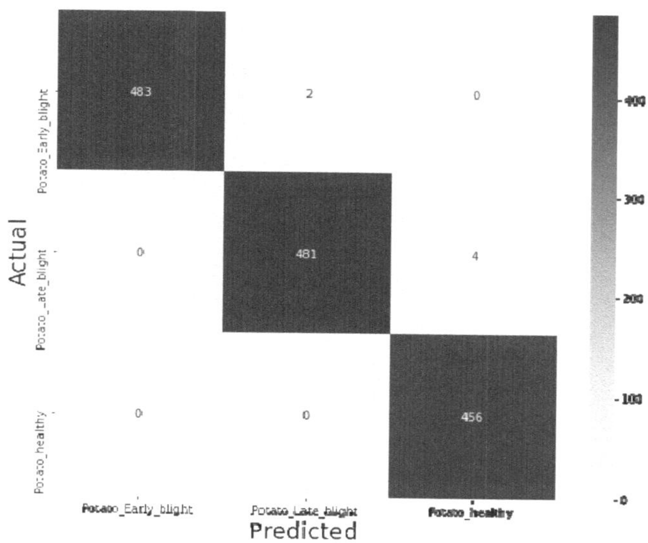

(c) VGG16 Model

Fig. 3: Confusion Matrix

In VGG16 for class 3 all images were classified correctly but for class 2 (late blight) out of 485 images 481 were classified correctly and 4 images were misclassified into class 3. For class 1 out of 485 images only 2 images were misclassified into class 2 rests all images were classified correctly.

CLASSIFICATION REPORT

Precision, recall (sensitivity) and F1 score generated for AlexNet, MobileNet and VGG16 is presented in Table 4 for three classes of potato leaves namely early blight, late blight and healthy class. In classification report it is shown that F1 score for AlexNet model for all three classes is 0.99, 0.99 and 1.00. In the same way F1 score for the MobilNet model for all three classes is 0.99, 0.99 and 1.00. Whereas, F1 score for VGG16 is 1.00, 0.99 and 1.00 for all three classes respectively. Sample images of early blight, healthy and late blight potato leaves are shown in Figure 4.

Table 4: Classification Report for AlexNet, MobileNet and VGG16 Deep CNN Models

	AlexNet			*MobileNet*			*VGG16*			*Support*
	Precision	*Recall*	*F1 Score*	*Precision*	*Recall*	*F1 score*	*Precision*	*Recall*	*F1 Score*	
Early blight	0.99	1.00	0.99	0.98	1.00	0.99	1.00	1.00	1.00	485
Late blight	1.00	0.99	0.99	1.00	0.98	0.99	1.00	0.99	0.99	485
Healthy	0.99	1.00	1.00	1.00	1.00	1.00	0.99	1.00	1.00	456

(A) Early Blight (b) Healthy (c) Late Blight

Fig. 4: Sample Images of Potato Leaves

MEMORY REQUIREMENT

Table 5 presents memory requirements for all three trained deep CNN VGG16, AlexNet and MobileNet models for potato disease detection. In order to deploy deep CNN models in edge nodes, model should occupy less memory because edge nodes have fewer resources like main memory, CPU and storage capacity in contrast to desktop and laptop. By observing the memory needs of these 3 deep CNN models, it is argued that small models like MobileNet would be best suited for edge nodes for real time disease detection.

Table 5: Memory Need for 3 Deep CNN Models

Deep CNN Models	*Memory (MB)*
VGG16	1024
AlexNet	374.3
MobileNet	22.5

CONCLUSION

In this work AlexNet, MobileNet and VGG16 three deep CNN models to classify potato leaves images and detect early blight, late blight and healthy plants were explored. The prime objective of this work is to identify a lightweight CNN model that can be deployed at edge node smoothly and analyses potato leaves images instantaneously. In order to load trained modified AlexNet, MobileNet and VGG16 models into edge node they require main memory (MB) 1024, 374.3 and 22.5 respectively. Results clearly indicate that the modified MobileNet model requires much less memory in contrast to AlexNet and VGG16 models so MobileNet is selected as a light-weighted CNN model to deploy into edge node for real time identification of diseases. The advantage of utilizing this approach is that there is no need to send samples of leaf images to the cloud or laboratory to analyse them. Despite being lightweight, MobileNet achieved the same accuracy which is 99.87% for both train and test images. In the case of MobileNet, results of F1 score obtained for class-1 (early blight), class-2 (late blight) and class-3 (healthy) is 0.99, 0.99 and 1.00 respectively, which is pretty good. Future works can be focussed on extending this work to various other crops like tomatoes, soybeans, rice and wheat.

REFERENCES

[1] Antonini, M., Vecchio, M. and Antonelli, F. (2019). Fog computing architectures: A reference for practitioners. IEEE Internet of Things Magazine, 2(3), 19–25.

[2] Bangui, H., Rakrak, S., Raghay, S. and Buhnova, B. (2018). Moving to the edge-cloud-of-things: Recent advances and future research directions. Electronics, 7(11), 309.

[3] Barbedo, J.G.A. (2016). A review on the main challenges in automatic plant disease identification based on visible range images. Biosystems Engineering, 144, 52–60.

[4] Bhattarai, S. (2018). New Plant Diseases Dataset. https://www.kaggle.com/vipoooool/new-plant-diseases-dataset

[5] Coakley, S.M., Scherm, H. and Chakraborty, S. (1999). Climate change and plant disease management. Annual Review of Phytopathology, 37(1), 399–426.

[6] Dasgupta, S.R., Rakshit, S., Mondal, D. and Kole, D.K. (2020). Detection of diseases in potato leaves using transfer learning. In Computational Intelligence in Pattern Recognition (pp. 675–684). Springer.

[7] Ferentinos, K.P. (2018). Deep learning models for plant disease detection and diagnosis. Computers and Electronics in Agriculture, 145, 311–318.

[8] Fuentes, A., Yoon, S., Youngki, H., Lee, Y. and Park, D.S. (2016). Characteristics of Tomato Plant Diseases—A study for tomato plant disease identification. Proc. Int. Symp. Inf. Technol. Converg, 1, 226–231.

[9] Henfling, J.W. (1987). Late blight of potato (Vol. 4). International Potato Center.

[10] Howard, A.G., Zhu, M., Chen, B., Kalenichenko, D., Wang, W., Weyand, T., Andreetto, M. and Adam, H. (2017). Mobilenets: Efficient convolutional neural networks for mobile vision applications. ArXiv Preprint ArXiv:1704.04861.

[11] Islam, F., Hoq, M.N. and Rahman, C.M. (2019). Application of Transfer Learning to Detect Potato Disease from Leaf Image. 2019 IEEE International Conference on Robotics, Automation, Artificial-Intelligence and Internet-of-Things (RAAICON), 127–130.

[12] Islam, M., Dinh, A., Wahid, K. and Bhowmik, P. (2017). Detection of potato diseases using image segmentation and multiclass support vector machine, 2017 IEEE 30th Canadian Conference on Electrical and Computer Engineering (CCECE), 1–4.

[13] Jadhav, S.B., Udupi, V.R. and Patil, S.B. (2020). Identification of plant diseases using convolutional neural networks. International Journal of Information Technology, 1–10.

[14] Kalvakolanu, A.T.S. (2020). Plant Disease Detection from Images. ArXiv Preprint ArXiv:2003.05379.

[15] Krizhevsky, A., Sutskever, I. and Hinton, G.E. (2012). Imagenet classification with deep convolutional neural networks. Advances in Neural Information Processing Systems, 25, 1097–1105.

[16] Limbore, N.V. and Khillare, S.K. (2015). An analytical study of Indian agriculture crop production and export with reference to wheat. Review of Research Journal, 4(6).

[17] Lu, Y., Yi, S., Zeng, N., Liu, Y. and Zhang, Y. (2017). Identification of rice diseases using deep convolutional neural networks. Neurocomputing, 267, 378–384.

[18] Mohanty, S.P., Hughes, D.P. and Salathé, M. (2016). Using deep learning for image-based plant disease detection. Frontiers in Plant Science, 7, 1419.

[19] Munyaneza, J.E., Crosslin, J.M., Buchman, J.L. and Sengoda, V.G. (2010). Susceptibility of different potato plant growth stages to purple top disease. American Journal of Potato Research, 87(1), 60–66.

[20] Nutter, F.W., Esker, P.D. and Netto, R.A.C. (2006). Disease assessment concepts and the advancements made in improving the accuracy and precision of plant disease data. European Journal of Plant Pathology, 115(1), 95–103.

[21] Rangarajan, A.K., Purushothaman, R. and Ramesh, A. (2018). Tomato crop disease classification using pre-trained deep learning algorithm. Procedia Computer Science, 133, 1040–1047.

[22] Singh, V. and Misra, A.K. (2017). Detection of plant leaf diseases using image segmentation and soft computing techniques. Information Processing in Agriculture, 4(1), 41–49.

[23] Zwolenski, M. and Weatherill, L. (2014). The digital universe: Rich data and the increasing value of the internet of things. Journal of Telecommunications and the Digital Economy, 2(3), 41–47.

Moonlight and Time Based Dynamic Power Control for Road Lighting

Madhuri Unde[1] and Abhishek Singh Rathore[2]

ABSTRACT

It's a matter of concern for everybody across the world how to optimize the limited energy resources. LED lights are used to minimize the energy expenditure. This paper depicts a simple, strong and energy efficient road light control framework. The guideline utilized is thought of regular light (moon light and daylight) intensity and time based info control of the road light. The astronomical clock contingent upon geographical territory is considered to produce data base. The input energy of road light varies according to the variations in natural light intensity obtained from the data base. After mid night the street activity is too less so framework enlightens lights with half intensity so this utilizes nearly half of the power. The framework utilizes real time clock, microcontroller and TRIAC based circuit for controlling input energy of a road light. This shows that the real time clock based framework is more reliable than sensor based frameworks.

Keywords: Control, Dimming, Firing Angle, RTC.

INTRODUCTION

It is mandatory to be watchful and measure utilization of light around evening time by focusing on when and where light is required. The roadside lighting practices, such as "quality lighting" puts less burden on the energy resources and gives impressive paybacks also. Absence of glare and excessive contrast brings better visibility (Retrieved from http://www.darksky.org). India has a limit of 237.742 GW. The total yearly generation is 1053.9 Tetra Watt-hours (TWH). The road lights consumes a lot of electrical energy. The Bureau of Energy Efficiency, in view of Central Electricity Authority insights, has assessed net energy utilization for road lighting to be 6,131 million kWh in India. Lighting can represent 10–38% of the aggregate electrical energy charge in urban areas. Energy capable strategies can lessen road lighting costs (Swathi *et al.*, 2014). Proper lighting is important to give our towns and urban communities a quality

[1] Research Scholar, Shri Vaishnav Institute of Information Technology, Shri Vaishnav Vidyapeeth Vishwavidyalaya, Indore, Madhya Pradesh, India. E-mail: madhuriunde123@gmail.com

[2] Associate Professor, Shri Vaishnav Institute of Information Technology, Shri Vaishnav Vidyapeeth Vishwavidyalaya, Indore, Madhya Pradesh, India. E-mail: abhishekatujjain@gmail.com

look after sunset and also to address to the security issues (Retrieved from http://www.darksky.org).

These days, road lighting control is majorly done through manual control, which can at times be misused by labour leading to wastage of energy. On the other hand utilizing intensity control strategy, by setting up control circuit, can change the voltage across light dependent resistors to control road lights. This turns on lights naturally at night after dark and turns it off automatically in the morning, yet it is not 100% reliable because at times night road lighting becomes too bright and is a misuse of electricity (Lianhao, 1997). The second is time based control technique (that is, every once in a while opening and shutting control) (Joshi *et al.*, 2013). The second technique for road light control accomplishes programmed control of road light. Along these lines lessening the light intensity can bring down work costs and enhance the effectiveness. There is no thought of the accessibility of moon light, if natural light is accessible then there is need to waste electrical energy on the road lights. The proposed approach portrays the utilization of moon light with the current road light control frameworks.

There are a few trials in which the electricity wastages over the road lights are diminished. A framework with movement and light sensor is utilized as a part of few spots to decrease energy consumption by road lights (Velaga *et al.*, 2012). Smart road lights speak with each other to control a road light on pedestrians. These brilliant streetlights trade data of detecting moving objects and turn on the lamps (Mullner *et al.*, 2011). Next, image processing system is additionally used to lessen electricity utilization of road lights; the technique proposed takes the pictures of road lights by cameras and controls power of road lights (Zanjani *et al.*, 2012). Lot of work has been done on streetlight control yet some investigations concentrate on keen road light control that carries on looking like regular road lights in view of sensors, time based on-off control and recently advanced control frameworks like PLCs are generally utilized in industries (Swathi *et al.*, 2014). Still numerous road light control frameworks require a lot of electrical power. The spending cost for the electricity is expensive besides the discharge of CO_2 causing natural contamination. The framework application in streetlight control for every light will bring down the power required, maintenance cost, and increment availability of road light (Veena *et al.*, 2016).

Our predecessors had a limited capacity to work amongst morning and night. The light from the flames of petroleum products broadened their usefulness during the night, especially in encased zones. Outside, an extra light source was typically made accessible to complete action during the evening which is moonlight. The moon gives its most extensive light during the full moon (Retrieved from http://www.illinoislighting.org/moonlight.html). In the very objective behind road light to use minimum power, higher output and superior capacity can be accomplished by LED lights. Here TRIAC dimmers can be utilized, which are getting to be plainly prominent over SCR dimmers due to their definitive features. Various methodologies have been used to LED driver which demonstrates similarity with TRIAC dimmer (Lukas *et al.,* 2015). It is reasonable to have a cost profitable framework that decreases frequent maintenance, with reliable data and permits vigorous control over the lighting. This can be accomplished by effective, strong and low power framework.

IMPLEMENTATION

Proposed System

Road light power over time can be accomplished by proposing a less expenditure normal light (daylight and moonlight) intensity based and time based approach utilizing microcontroller. The measurable patterns of varieties in dusk, dawn and moonlight intensity during the year through various seasons and a scope of power necessities everywhere throughout the day are considered.

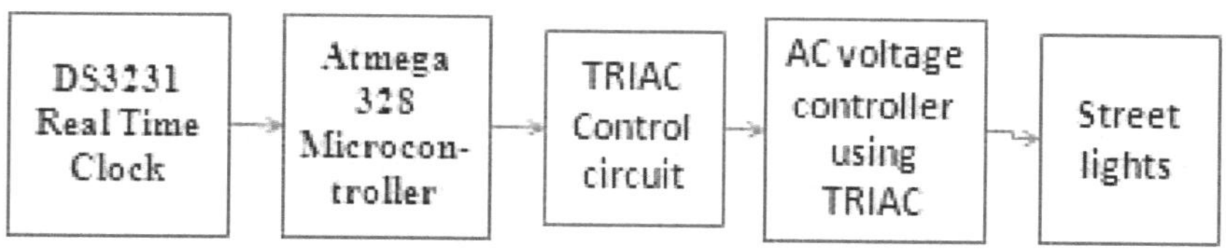

Fig. 1: Block Diagram of a System

The DS3231 Real Time Clock (RTC) gives time and date month and year data to the microcontroller. The micro-controller compares the information from RTC and data base and afterwards settle the required firing angle of TRIAC circuit as shown in Figure 1. Data base comprises of the everyday timings of nightfall, dawn and moon light intensity patterns during different seasons and then software program is written to decide the firing angle of AC voltage controller.

Hardware Implementation

The RTC IC DS3231 shares data with microcontroller Atmega328 by means of I2C serial communication interface. Microcontroller gets information from RTC IC and produces high frequency pulses for the TRIAC with firing angle in view of required light power. The firing pulses generated by microcontroller are then given to TRIAC control circuit, which comprises of transistor and isolation transformer. Pulses are given to the base of transistor and transistor is connected to the primary of isolation transformer. The need of isolation transformer is to give electrical isolation between low power control circuit which is the microcontroller and high power circuit which is only AC voltage controller utilizing TRIAC is as appeared in Figure 2.

The RTC DS3231 works faithfully between 0°C to +70°C. It is a low-cost, extremely accurate I2C real-time clock (RTC) with an integrated temperature compensated crystal oscillator (TCXO) and crystal.

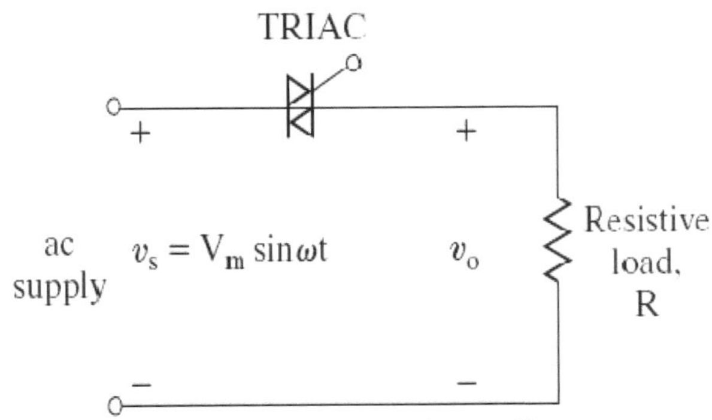

Fig. 2: AC Voltage Controller

Software Implementation

The intensity pattern throughout the year of natural light (sunlight and moonlight) has been considered and for corresponding light intensity firing angle of TRIAC control circuit is decided and after mid night system illuminates lights with 50% intensity so for this firing angle of TRIAC control circuit is 60 degree. Table 1 shows the light levels at Full Moon on a clear night and other days, from this data base the microcontroller can decide input power of street lights, so that it can produce required luminance. For example on full moon night, system controls the lamp intensity at 50% of its full intensity value. Similarly on dark night, system controls the lamp intensity which at 100% of its full intensity value. Table 3 gives the idea about moonrise/moonset timings as well as moon light illumination. Similar type of information means the moon calendar of three years is used for creating data base.

Figure 2 shows yearly sun graph for New Delhi city, this gives sunrise/sunset timings, these timings are considered and data base of Table 2 is created. About the data base of Table 2, the ON time of street light is 20 minutes late than sunset time similarly the OFF time of street light is 20 before than sunrise time, this is because just before sunrise and just after sunset there is sufficient light available. Software program takes the input data from RTC i.e. current time, date, month and year. Initially it compares this with the data base i.e. with Table 2 and microcontroller turns ON/OFF the street lights. Once light get turn on then second task of software is to compare current timings with data base which is prepared from moonset/moonrise timings then it generates appropriate firing angle for TRIAC control circuit. So that street light provides required intensity. Third task of software is to deem the lamps by half intensity after midnight.

Table 1: Moonlight Illuminance Retrieved from www.timeanddate.com/astronomy/india/new-delhi

Illuminance	*Day*	*Street Light Intensity*
0.25 lux	Full Moon on a clear night	60%
0.01 lux	Quarter Moon	90%
0.002 lux	Starlight clear moonless night sky including airglow	100%
0.0002 lux	Starlight clear moonless night sky excluding airglow	100%
0.00014 lux	Venus at brightest	100%

Table 2: ON OFF Timings of Street Light

Month	*Sunset*	*ON Time*	*Sunrise*	*OFF Time*
January	18:11	18:31	6:13	5:57
February	18:31	18:51	7:14	6:54
March	18:43	19:03	6:59	6:39
April	18:51	19:11	6:34	6:14
May	19	19:20	6:12	5:52
June	19:11	19:31	6:02	5:42
July	19:19	19:33	6:06	5:46
August	19:13	19:30	6:17	5:57
September	18:52	19:12	6:25	6:05
October	18:26	18:46	6:30	6:10
November	18:04	18:24	6:40	6:20
December	17:59	18:19	6:57	6:37

The algorithm for the software is given by,

1. Initialize I2C communication module.
2. Initialize RTC DS3231.
3. Initialize clock system.
4. While (1).
5. Get current time, date, month and year from RTC.
6. Compare month with data base and turn ON/OFF lights as per the timings mentioned in Table 2.
7. Once again compare date with moon calendar and control street light intensity according to Table 1.
8. Keep the full intensity at dark night till 23:59 pm and after mid night keep the 50% intensity till 4 am.
9. During 4 am to 20 minutes before sunrise keep 80% intensity.
10. And turn off the lamps at 20 minutes before sunrise.

Lightnig an Empty Street

As shown in Figure 3 a vacant road is lighted with 100% light intensity, thus this is the wastage of enormous power. In this manner after 12 pm when the street activity is too less, the proposed framework highlights that with half intensity and investment on power can be reduced.

Fig. 3: Empty Streets

Table 3: Moonrise, Moonset and Moonlight Illuminance for the May 2017 Retrieved from www.timeanddate.com/moon/india/new-delhi

2017	Moonrise/Moonset			Meridian Passing		
May	Moonrise	Moonset	Moonrise	Time	Distance (km)	Illumination
2	-	00:07 ↖ (289°)	11:44 ↗ (72°)	18:23 (87.9°)	375,186	43.8%
◐ 3	-	01:00 ↖ (287°)	12:42 ↗ (74°)	19:16 (85.1°)	380,450	55.1%
4	-	01:48 ↖ (284°)	13:38 ↗ (78°)	20:06 (81.6°)	385,513	65.8%
5	-	02:32 ← (280°)	14:31 → (82°)	20:54 (77.7°)	390,160	75.5%
6	-	03:13 ← (276°)	15:23 → (86°)	21:39 (73.4°)	394,271	83.9%
7	-	03:52 ← (272°)	16:13 → (91°)	22:23 (69.2°)	397,795	90.6%
8	-	04:30 ← (267°)	17:03 → (95°)	23:07 (65.0°)	400,721	95.6%
9	-	05:08 ← (263°)	17:52 → (99°)	23:51 (61.2°)	403,051	98.6%
10	-	05:47 ← (259°)	18:42 ↘ (103°)	Moon does not pass the meridian on this day.		
○ 11	-	06:27 ↙ (256°)	19:31 ↘ (106°)	00:35 (57.8°)	404,772	99.8%
12	-	07:08 ↙ (253°)	20:21 ↘ (108°)	01:20 (55.0°)	405,848	99.0%
13	-	07:52 ↙ (251°)	21:10 ↘ (110°)	02:07 (52.9°)	406,209	96.4%
14	-	08:37 ↙ (250°)	21:59 ↘ (110°)	02:54 (51.6°)	405,764	92.0%
15	-	09:25 ↙ (250°)	22:46 ↘ (110°)	03:42 (51.2°)	404,409	86.0%
16	-	10:14 ↙ (250°)	23:32 ↘ (109°)	04:30 (51.7°)	402,056	78.5%
17	-	11:05 ↙ (252°)	-	05:18 (53.2°)	398,653	69.8%
18	00:17 ↘ (107°)	11:57 ↙ (255°)	-	06:06 (55.5°)	394,222	60.1%
◑ 19	01:01 ↘ (104°)	12:49 ↙ (258°)	-	06:54 (58.6°)	388,875	49.8%
20	01:44 → (100°)	13:44 ← (262°)	-	07:42 (62.5°)	382,840	39.0%

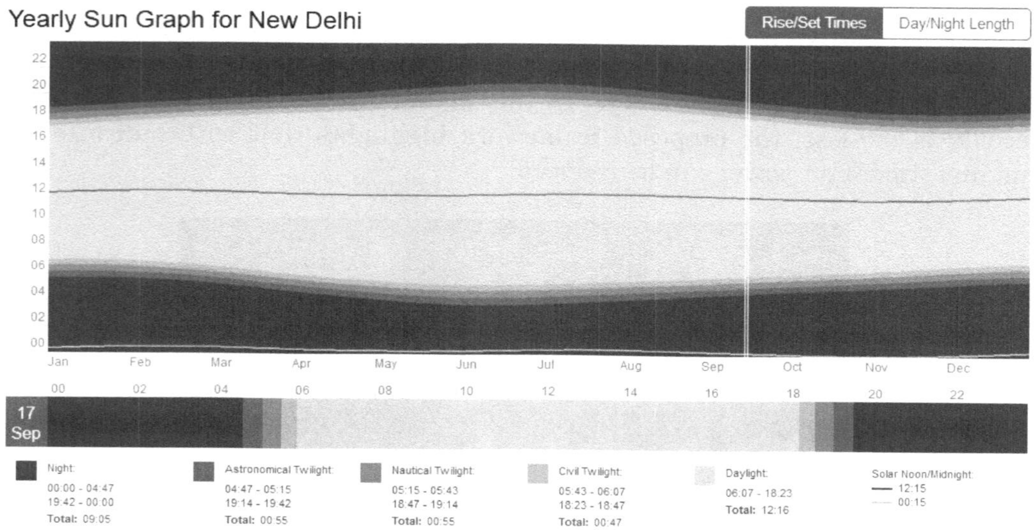

Fig. 4: Yearly Sun Graph for New Delhi

Illumination Calculation

The Illumination can be calculated as,

$$I = Ll\ Cu\ LLF/Al. \quad \ldots (1)$$

Where

I = illumination

Ll = lumens per lamp

Cu = utilization coefficient

LLF = light loss factor

Al = area per lamp (m^2)

Example

5 incandescent lamps of *500 W (10600 lumens per lamp)* are used in an area of *50 m*^{2}. With *Cu = 0.6* and *LLF = 0.8* illumination can be calculated as

$$I = 5\ (10600 \text{ lumens})\ (0.6)\ (0.8)/(50 \text{ m}^2) = 509 \text{ lux.}$$

Phase Angle Detection Using Microcontroller

TRIAC has three terminals specifically MT1, MT2 and gate. During every half cycle of AC, a gate pulse is required to turn on. Microcontroller pin can drive TRIAC gate. The waveform across load is appeared in Figure 4. RTC feed date; time and month to microcontroller, in view of this and moon light intensity, high frequency pulses are generated with proper firing angle. TRIAC required that firing pulses for gate driving circuit. In the framework isolation transformer is utilized as a part of gate driving circuit for isolation reason.

RESULTS

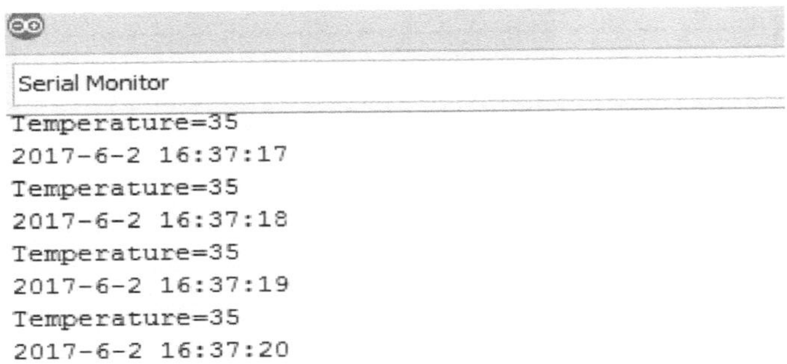

Fig. 5: RTC Readings on Serial Monitor

Figure 5 shows the serial monitor of programming tool, over which the data from RTC IC DS3231 is received. It gives the current time in seconds, minutes and hours as well as RTC sends date information with month and year and temperature to microcontroller.

Table 4: Load Voltage, Load Current and Firing Angle Values

Sr. No.	*Firing Angle(α) in Degree*	*Voltage Across Load (V)*	*Load Current (A)*	*Power Consumption (W)*
1.	0	230	0.17	39.1
2.	30	224	0.16	35.84
3.	60	204	0.15	30.6
4.	90	160	0.11	17.6
5.	120	100	0.07	7
6.	150	37	0.028	1.03
7.	180	0	0	0

As observed from the Table 4 and Figure 7, power consumed decreases significantly with increasing firing angle. The variation of firing angle is programmed to occur at a step of 30 degrees. The approximate energy saved per day per street lamp is to 30.02 WHr. Taking into consideration the quantity of road lights being utilized the aggregate sparing would be in thousands of MWHr. Figure 6 is demonstrating the input mains waveform and load voltage waveform with firing angle (α).

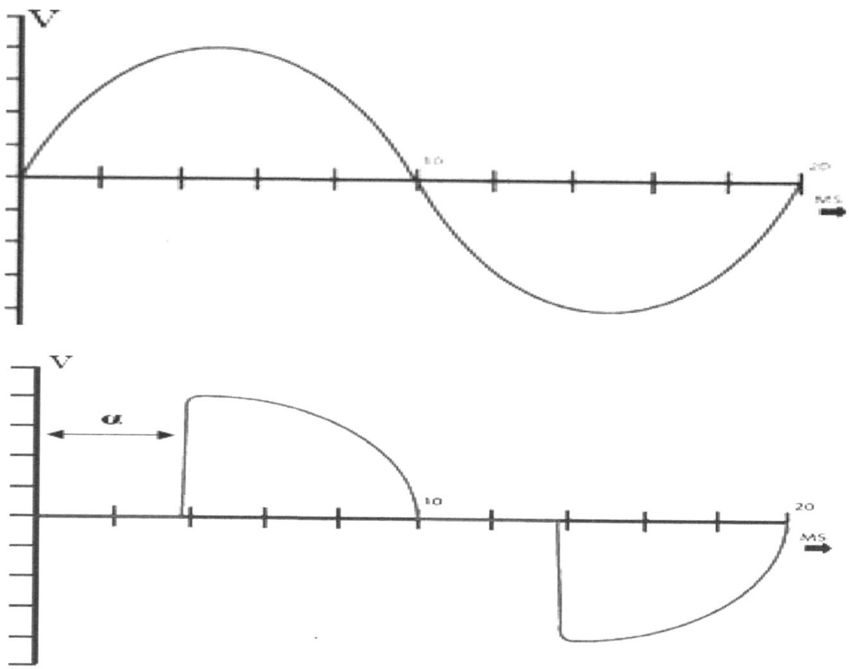

Fig. 6: Mains Supply Waveform and Waveform across Load

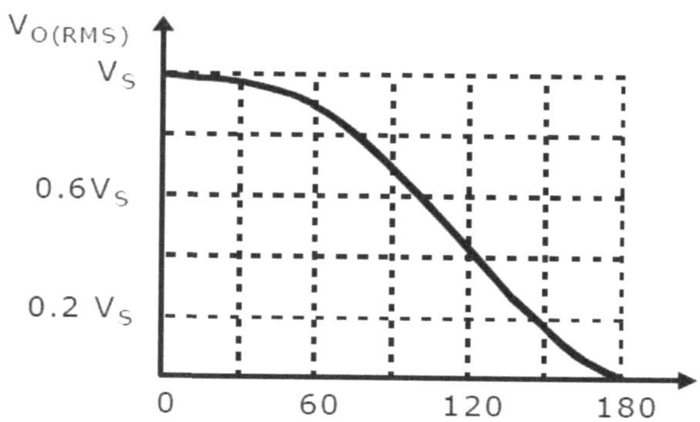

Fig. 7: Load Voltage vs Firing Angle in Degree

CONCLUSION

Dynamic intensity control exhibited in the proposed framework brings about the effectiveness of the Quality lighting system. For accomplishing high effectiveness, expanded existence of lights and lessening in CO_2 discharge, utilization of daylight and additionally moon light is essential. Proposed framework is the best answer for the issues like paying extra cost and minimize the power usage. Proposed work gives a stage to the different diminishing innovations.

ACKNOWLEDGMENT

The authors of this paper are very much thankful of Sanjivani College of Engineering, Kopargaon for their support and for providing the best infrastructure.

REFERENCES

[1] Bright as The Full Moon: How Much to Light Up The Night? A reprint of the Illinois Coalition for Responsible Outdoor Lighting website page at http://www.illinoislighting.org/moonlight.html.

[2] Chun-An Cheng, Chien-Hsuan Chang, and Tsung-Yuan Chung (2015). "Design and Implementation of a Single-Stage Driver for supplying an LED Street Lighting Module with power factor corrections," *IEEE Trans. Power Electron.*, Vol. 30, No. 2.

[3] Dongkyung, Park; Zhidong, Liu and Hoi, Lee (2014). "0 V 10 W 93%-Efficiency Current-Accuracy-Enhanced Dimmable LED Driver with Adaptive Timing Difference Compensation for Solid-State Lighting Applications.," *IEEE journal of Solid-state-circuits*, Vol. 49, No. 8.

[4] LIU Lianhao (1997). "A new street lamp controller design" (J). Computing Technology and Automation, (4): 61–63.

[5] Lukas, Lohaus; Arne, Rossius; Stefan, Dietrich; Ralf, Wunderlich and Stefan, Heinen (2015). "A Dimmable LED Driver with resistive DAC feedback control for adoptive voltage regulation," *IEEE Trans. Industrial application,* Vol. 51, No. 4.

[6] Mukul, Joshi; Rajashri, Madri; Shruti, Sonawane and Abhinav, Gunjal (2013). "On Time Based Intensity Control for Energy Optimization Used for Street Lighting," *IEEE Texas Instruments India Educators' Conference*, pp. 211–215.

[7] Mullner, R., Riener, A.A. (2011). "An energy efficient pedestrian aware Smart Street Lighting system", *International Journal of Pervasive Computing and Communications*, Vol. 7, Issue 2, pp. 147–161.

[8] Outdoor lighting code handbook, Version 1.14, December 2000/September 2002, International Dark-Sky Association, http://www.darksky.org

[9] Zhang, R. and Chung, S.H. (2014). "A TRIAC dimmable LED lamp driver with wide dimming range," *IEEE Trans. Power Electron.*, Vol. 29, No. 3, pp. 1434–14446.

[10] SangCheol, Moon; Gwan-Bon, Koo and Gun-Woo, Moon (2015). "Dimming-Feedback control method for TRIAC Dimmable LED drivers," *IEEE Trans. Industrial Electron.*, Vol. 62, No. 2.

[11] Swathi, C.A., Hemanth Kumar, S. and Annappa, A.R. (2014). "Smart Street Lighting System Based on Sensors Using PLC and SCADA", *IJMET,* Vol. 5, Issue 9, September, pp. 44–50.

[12] Veena, P.C., Paulsy, Tharakan; Hima, Haridas; Ramya, K.; Riya, Joju and Jyothis, T.S. (2016). "Smart Street Light System based on Image Processing", *IEEE Conference on Circuit, Power and Computing Technologies [ICCPCT]*

[13] Velaga, R. and Kumar, A. (2012). "Techno-economic evaluation of the feasibility of a smart street system: A case study of rural India", *Procedia Social and Behavioral Sciences,* Vol. 62, pp. 1220–1224.

[14] www.timeanddate.com/astronomy/india/new-delhi

[15] www.timeanddate.com/moon/india/new-delhi

[16] Zanjani, P.N., Ghods, V. and Bahadori, M. (2012). "Monitoring and remote sensing of the street lighting system usingcomputer vision and image processing techniques for the purpose of mechanized blackouts", *Proceedings of 19th International Conference on Mechatronics and Machine Vision in Practice,* pp. 5–10.

Presumptive Study of Analysis of Cannabis Metabolite in Pubic Hair of Drug Abuser

Dinesh Kamble[1] **and Kavita Sharma**[2]

ABSTRACT

Cannabis substance abuse is widely causing drug addiction problems across the world. In the drug abuse cases, hair plays an important role in identifying drug abuse and drug history consumption of the addict. However, scalp hair may not be available in every drug abuse cases, but hair of other body parts play an essential role in forensic toxicological analysis. This study attempts to presumptive identification of cannabis metabolite in pubic hair using Fast Blue B Salt Test.

Keywords: Drug Abuser, Fast Blue B Test, Forensic Toxicology, Pubic Hair.

INTRODUCTION

Some of the common illicit drugs are Hashish, Marijuana, Ganja, and Bhang, in which the psychoactive products are procured from *Cannabis sativa* plants (Grotenhermen, 2007). Cannabis is the most widely produced and abused drug across the world. About 192 million people consume cannabis. Its consumption is increasing in developing countries as compared to that in developed countries. In the world, more than half of drug offense cases reported from 2014 to 2018 were due to cannabis (https://wdr.unodc.org/wdr2020). India is a developing country, and about 2.8% of Indians aged 10–75 years (3.1 crore individuals) are currently using cannabis (Ambekar *et al.*, 2019). The identification of cannabis in viscera or other biological fluid is a challenge to forensic science. Therefore, hair is a broader window for drug abuse cases to extract and confirm cannabis metabolites for a longer duration. In every drug abuse case, scalp hair is not available for forensic toxicological analysis. In this scenario, body hair plays an essential role in forensic analysis. The drug incorporation in hair through the passive diffusion model is proposed (Manolis *et al.*, 2017 & Offidani C. *et al.*, 1992), which opens the window for hair analysis in case of drug-related deaths. Hair plays an important

1 Research Scholar, Shri Vaishnav Institute of Forensic Science, Shri Vaishnav Vidyapeeth Vishwavidyalaya, Indore, Madhya Pradesh, India. E-mail: dineshkamble@svvv.edu.in

2 Professor, Shri Vaishnav Institute of Science, Shri Vaishnav Vidyapeeth Vishwavidyalaya, Indore, Madhya Pradesh, India. E-mail: kavitasharma5872@gmail.com

role in identifying drug abuser history even after single-dose exposure. It is possible to detect hair samples with non-invasive techniques (Henderson, 1993). In hair, drug metabolites are incorporated and deposited for which the detection is made through metabolite identification.

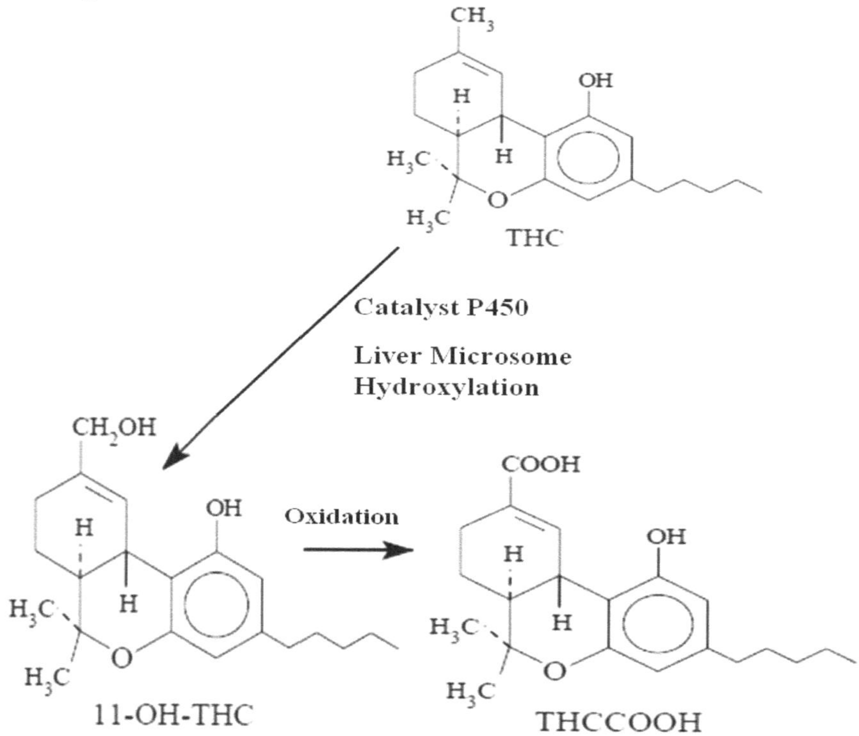

Cannabis is converted into 11-hydroxy-Δ9-tetrahydrocannabinol (11-OH-THC) by hydroxylation and further converted into 11-nor-9-carboxy-Δ9-tetrahydrocannabinol (THCCOOH)

Fig. 1: Mechanism of Cannabis in Forming THCCOOH

MATERIALS AND METHOD

Chemicals and Reagents

11-Nor-9-carboxy-Δ^9-Tetrahydrocannabinol is used as a standard which was purchased from Cerilliant TX, USA. Methanol, Double distilled water of analytical reagent grade, and Dimethyl sulfoxide solvent was used.

Fast Blue B Salt Test

1. *Reagent A:* Petroleum Ether.
2. *Reagent B:* Fast Blue B Salt (1% w/w diluted with anhydrous sulphate).
3. *Reagent C:* Sodium Bicarbonate (10% aqueous solution).

Procedure

The few drops of pulverized sample put into the center of the filter paper. Two drops of reagent A were added, allowing the liquid to penetrate to the filter paper. Very small amount of reagent B was added to the center of the filter paper and then two drops of reagent C added.

Sample Collection

Pubic hair of cannabis abuser was collected from Gwaliar Mansik Arogyashala Centre, Gwalior, Madhya Pradesh, India. The subject is a 10-year old cannabis abuser who discontinued taking cannabis. The pubic hair sample was collected after 40 days of cannabis usage. The consent was taken from rehabilitation centre and subject before collecting the samples.

Sample Inclusion Criteria

- Pubic hair: only cannabis abuser.

Sample Exclusion Criteria

- The subject has not taken cannabis along with other psychoactive drugs.
- If the subject has abnormality related to health.
- The subject was observed for any psychotic disorder.

Sample Preparation

50 mg sample was taken using an electronic balance and washed using double distilled water twice for 5 minutes. The washed hair sample was dried using an oven at 60–70 degree for 2 minutes. The dried sample was cut into small pieces and dissolved in 2 ml of methanol in a beaker, sonicated for 3 hours, and the same procedure was repeated (Boldis *et al.*, 2003). After sonicated, the extract was filtered and decanted. The extracted residue took spot on filter paper and standard spot on another filter paper.

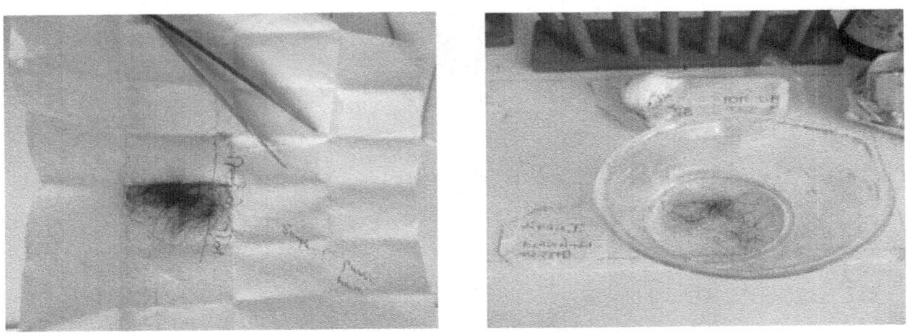

Fig. 2: Public Hair Sample

Fig. 3: Extraction and Sample Preparation

RESULTS AND DISCUSSION

The extracted residue obtained from cannabis metabolites and that residue has been poured on filter paper and added Fast Blue B Salt Test reagent. Its shows purple red coloration and purple red coloration indicates the presence of cannabis metabolites in the sample. The intensity of coloration differs because of the concentration of the interested compound in the samples are different.

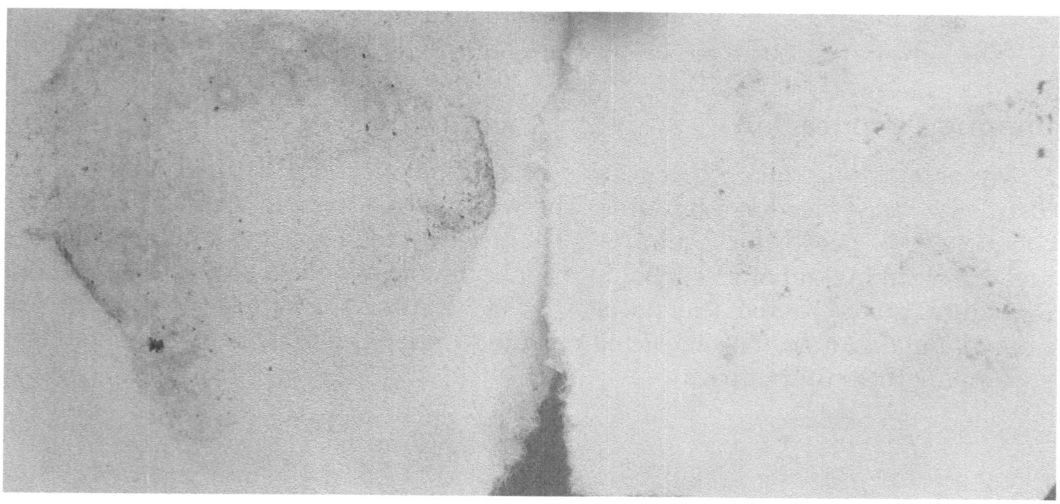

Standard Sample

Fig. 4: Reaction of Extracted Metabolites of Cannabis Metabolites with Fast Blue B Salt Test Reagent

Conflicts of Interest

The authors declare that they have no conflicts of interest.

REFERENCES

[1] Ambekar, A., Agrawal, A., Rao, R., Mishra, A.K., Khandelwal, S.K. and Chadda, R.K. on behalf of the group of investigators for the National Survey on Extent and Pattern of Substance Use in India (2019). Magnitude of Substance Use in India. New Delhi: Ministry of Social Justice and Empowerment, Government of India. http://socialjustice.nic.in/writereaddata/UploadFile/Survey%20Report.pdf

[2] Boldis, O., Kocsis, Gy, Gachalyi A. and Furesz, J. (2003). Gas Chromatography-Mass Spectroscopy-Single Ion Monitoring Measurement of 11-THC-COOH in Urine. *Journal of Chromatographic Science*, Vol. 41.

[3] Grotenhermen, F. (2007). The toxicology of cannabis and cannabis prohibition, *Chem Biodivers,* 4(8): 1744–69.

[4] Henderson, G.L. (1993). Mechanism of Drug Incorporation into Hair, *Forensic Science International*, Vol. 63: 19–29.

[5] Manolis, N., Athanasios, K., Matthaios, P., Elena, V., Polychronis, D., Katerina, K., Alexander, I., Emmanouil, G. and Aristidis, M. (2017). Comparative Evaluation of Drug Depositionin Hair Samples Collected from Different Anatomical Body Sites, *Journal of Analytical Toxicology*, Vol. 41:214–223.

[6] Offidani, C., Strano, R. and Chiarotti, M. (1992). Drug Distribution in the Head, Axillary and Pubic Hair of Chronic Addicts, *Forensic Science International*, Vol. 63: 108–108.

[7] United Nations Office on Drugs and Crime (2020). World Drug Report: Global Drug Use Rising; While COVID-19 Has far Reaching Impact on Global Drug Markets, https://wdr.unodc.org/wdr2020.

Professional Skills in Chemistry through Perspectives of Industry, Research and Academia

Kiran Badave[1] and Shankar Thopate[2]

ABSTRACT

Basic Science is an important branch for the scientific development of a country. The scientific development of a country depends upon the quality of education. Basic Science students have wide opportunities in academia, research and industry. Domain knowledge along with the professional skills are required skills for the Science profession. India Skill report concluded that about 46.21% of students are employable. All the employer sector i.e. research, academia and industry are reluctant to absorb freshers since it takes time and money to make a fresher competent and skilled professional. Experienced candidate is preferred over a fresh candidate. Considering this fact academic institutions have taken initiatives to provide skill development education through Skill development programs through the programs of National Skill Development Corporation (NSDC) such as B.Voc. Certificate, Diploma and Degree courses and institutional level skill based courses.

Keywords: Academia, Chemistry, Industry, Research, Skills.

INTRODUCTION

Basic science has great capacity to alter the scenario of the world. Innovations such as electricity, telephone, vaccines, antibiotics etc. have changed our lives (Zohuri, 2016; Tom *et al.*, 1997; Rappuoli *et al.*, 2014; Adedeji, 2016). There is complete dependence on these innovations. However, the field may not fascinate many students, because of lack of awareness regarding the field. They perceive it as a long journey to achieve success in terms of monetary gains as compared to applied ones. According to the Economics Discussion, 2021 and India Skills Report, 2019, education has reached to the masses from primary to graduate level. The government is taking essential efforts to expand education facilities by setting up

[1] Research Scholar, Research Centre in Chemistry, B.P.H.E., Ahmednagar College, Ahmednagar (MS), India. E-mail: kiranbadave@gmail.com

[2] Principal, S.S.G.M. Science, Gautam Arts and Sanjivani Commerce College, Kopargaon (MS), India. E-mail: srthopate@gmail.com

higher education institutes, and providing funds for quality education. The funds for research, development of laboratories, upgradation of laboratories and purchase of equipment is sanctioned by the government for the research projects. In spite of all these efforts most of students are found to be ineligible for jobs as they lack professional skills and some are not able to sustain in their jobs. India Skill report concluded that about only 46.21% of students are employable (PMKY Guideline report, 2021).

Industry, academia and research are available sectors for basic science students. The skill based education is the need of hour, and will remain for future generations to come (Russell, 2017). As the world is in constant development mode the technology has changed the scenario. The incorporation of artificial intelligence will change future jobs opportunities (Bass, 1999). Research and development in all the sectors is essential for survival (Hellgren *et al.*, 1999). There is competition for excellence in all the sectors which creates work pressure to perform. Industry needs people with skills and wants returns on investments (Davies *et al.*, 2001; Stott *et al.*, 2012). Industries are reluctant to train people for jobs, but how a person gets professional experience without a job. There are lot of differences between working patterns in academia and industry. (Shams *et al.*, 2018). The national and private research institutes, industries and university affiliated teachers having research projects in some extent decreases this gap by providing opportunity to work in summer for research projects. But not everyone gets this opportunity. The academia, research and industry have to work hand in hand for empowering youth ultimately for the development of the country (Breaugh *et al.*, 2000).

Domain Knowledge

Knowledge is power (Perry, 2004). There is addition of knowledge in every single second. But, the fundamentals remain the same. Students need to clarify the fundamentals during their graduation. However, it is observed that students are afraid to ask questions (Stevens, 2005). This is a personal barrier that needs to be overcome by them or special efforts need to be taken by teachers to remove the barrier. Students need to be good in special subjects along with allied one such as physics, chemistry, mathematics, life sciences and geography too. Along with this student must be good in language for communication and writing i.e. documentation or report writing (Towns *et al.*, 2015). Students can add extra skill by learning foreign languages.

Laboratory Skills

Laboratory skills are highly appreciated by recruits since they are required for work (D'abate *et al.*, 2009). It is sometimes observed that students with average marks are seen with remarkable practical skills, while in some cases the students with excellent knowledge are below average in practical skills. Though the syllabus is framed keeping both the theory and practicals in mind. The practical and their conduction needs to be modified by academia. It's not the results of the practical but the involvement during practicals in the laboratory is important for good output. Students to teacher ratio needs to be minimized at the graduate level and more emphasis need to be given to student teacher interaction along with the role of the teacher as mentor. So there must be some teachers who are having professional experience in the research and industry. Some institutes need to take initiatives to keep faculty updated for students to understand the importance of college level learning, industry. Research oriented teaching and internships should be provided once a year for a week to give remarkable results (Tamir *et al.*, 1992).

It is observed that students are not sincere for gaining practical knowledge. They are very poor in knowledge about apparatus, their use keeping and learning methods, instruments the same. Calibration and use of equipment for solving the research problem analytically and also the knowledge about standard operating procedures. Use of personal protective equipment, use of fire extinguishers should be incorporated in the curriculum. The first step for the faculty with professional knowledge and enthusiasm is to provide this knowledge is the ultimate way to bridge the gap. The traditional practical methods develop the knowledge and skills while conducting the experiments. Today most students carry out initial set experiments. Hence all the remarkable or milestone steps are provided in the procedure. Laboratory practical textbooks renders the referencing ability of faculty and students (Connor-Greene, 2000). Most of the practicals today are cook-book experiments. The demands are different and training is given in different way and this creates the gap. They have to be guided to set up and execute the experiment and present the results and not be helped to obtain results and complete experiments. New experiments or research requires overall experiment monitoring.

Journal writing is also an important skill. Data collection is helpful for future planning and IPR applications (Chadwick *et al.*, 2018). So, experiments along with simultaneous documentation are equally important for any institution. The laboratory journal is a record of the experiments and the observations made along with the experiment results. During this journal writing method enables students to find the compound's molecular weight, physical constants such as melting point, boiling point, density etc. from authentic sources such as Merck-Index, Lange's Handbook. Calculating the moles and equivalent ratio in which quantity should be taken. Students also know the important information given on packing such as date of manufacturing, expiry date, batch no production or marketing agency, trademark and mainly safety symbols. In research papers the quantity of compound is given in moles not in gm or mg. This is new to even well qualified postgraduate students as this has not been in practice. Till the last year of post-graduate, the students are not trained to such practices, because textbooks are used over reference books. The journal writing is highly important in research and industry but this is taken casually in academia. Students do experiments and when in the following week they submit a journal for correction, it is not duly checked on time. Thereby the students do not develop important professional skills required for journal writing (Prosser *et al.*, 2007).

Observational ability is developed as they need to write every observational change such as exothermic, change in color or state of chemicals. Students need to write each and every progressive observation at definite along with mentioning time. The audits or IPR are important aspects in the professional world. So, documentation gets more importance. Style of writing is totally different as compared to college journal writing. Students must be aware before joining services about required scientific vocabulary, calculations methods, overwriting correction procedure and overall pattern of writing journal, useful academic sources. In academics one follows reference books or research papers or thesis, but in profession one needs to follow accrediting or auditing organization guidelines. This is an academic and professional gap. Hence, this method is the initial step towards the incorporating professional skill among the students.

The evidence of success will be seen among students themselves reading research papers or articles, referencing skills, journal writing and obtaining data, interpretation, observed results and presentation such students are performing well in their profession. Students feel confident while working (Roberts *et al.*, 2003). This enhances their self-esteem which is highly important for any fresher. They are benefited by this journal writing style. Students receive feedback from their supervisor and if they have completed any project work in the past and that becomes the ultimate reward for the teacher too.

Academia are not funded like research institutes and industry so students are ready for direct absorption. Government institutes such as DST, DBT, AICTE and UGC along with other granting agencies have taken initiatives for decreasing this gap. The results were seen in recruitment and development in research. Research institute is also taking initiatives providing summer project training along with an open day to showcase the research in the institutes. Research scholars also collaborate with academic fellows and are available as experts in seminars and conferences. Eminent personalities as research scientists are taken on higher position as emeritus professors or scientists at institutional or university level. Their experience is valuable for academia to set up an incubation and innovation center.

The expectations of recruiters are increasing day by day and to meet this challenge, student also need to focus on personality development (Kinnick *et al.*,

2005). Emotional intelligence needs to be at a high level at the workplace. The management at 360 degrees is inculcated in students in terms of time, health, money, relationship and society. The communication skills with all level personalities must be known so as to maintain dignity of themselves and others (Lucas, 1993). To keep work oriented and not be a problematic employee, continuous learning attitude must be developed (Gephart *et al.*, 1996). Polite, knowledgeable and skilled professionals is in demand in the society and the organizations.

CONCLUSION

The basic science will remain the need of the world always. The multidisciplinary work approach needs to be enhanced amongst the students. Industry academia and research institutes have to work hand in hand for the development of the country. Domain knowledge along with professional and personality development is a need of the present day society. Faculty with domain knowledge and professional skills is the only way to bridge the gap. Government and institutes should take initiatives for developing skilled manpower.

REFERENCES

[1] Adedeji, W.A. (2016). The Treasure Called Antibiotics. Annals of Ibadan post-graduate medicine, 14(2), 56–57. Retrieved 21 February 2021, from https://ugcbeta.inflibnet.ac.in/pdfnews/2519852_UDI10.pdf.

[2] Bass, B. (1999). Two Decades of Research and Development in Transformational Leadership. *European Journal of Work and Organizational Psychology 8*(1), 9–32. https://doi.org/10.1080/135943299398410

[3] Breaugh, J. and Starke, M. (2000). Research on Employee Recruitment: So Many Studies, So Many Remaining Questions. *Journal of Management, 26*(3), 405–434. https://doi.org/10.1177/014920630002600303

[4] Chadwick, S., de la Hunty, M. and Baker, A. (2018). Developing Awareness of Professional Behaviors and Skills in the First-Year Chemistry Laboratory. *Journal of Chemical Education 95*(6), 947–953. https://doi.org/10.1021/acs.jchemed.8b00025

[5] Connor-Greene, P. (2000). Making Connections: Evaluating the Effectiveness of Journal Writing in Enhancing Student Learning. *Teaching of Psychology 27*(1), 44–46. https://doi.org/10.1207/s15328023top2701_10.

[6] D'abate, C., Youndt, M. and Wenzel, K. (2009). Making the Most of an Internship: An Empirical Study of Internship Satisfaction. *Academy Of Management Learning & Education 8*(4), 527–539. https://doi.org/10.5465/amle.8.4.zqr527

[7] Davies, D., Taylor, R. and Savery, L. (2001). The role of appraisal, remuneration and training in improving staff relations in the Western Australian accommodation industry: a comparative study. *Journal of European Industrial Training 25*(7), 366–373. https://doi.org/10.1108/eum0000000005837

[8] Development of Education in India after Independence. *Economics Discussion.* Retrieved 21 February 2021, from https://www.economicsdiscussion.net/articles/development-of-education-in-india-after-independence/2293.

[9] Gephart, Martha A. (1996). Learning organizations come alive, Training & Development 50, 34–45.

[10] Govt. of India (2019). India Skills Report finds 46.21% of students employable. https://www.thehindubusinessline.com/news/education/india-skills-report-finds-4621-of-students-employable/article30269722.ece

[11] Govt. of India (2021). PMKY Guideline report PMKY Guideline report.

[12] Hellgren, J., Sverke, M. and Isaksson, K. (1999). A Two-dimensional Approach to Job Insecurity: Consequences for Employee Attitudes and Well-being. *European Journal of Work and Organizational Psychology 8*(2), 179–195. https://doi.org/10.1080/135943299398311

[13] Kinnick, K. and Parton, S. (2005). Workplace Communication. *Business Communication Quarterly 68*(4), 429–456. https://doi.org/10.1177/1080569905282099

[14] Lucas, M. (1993). A validation of types of career indecision at a counseling center. *Journal of Counseling Psychology 40*(4), 440–446. https://doi.org/10.1037/0022-0167.40.4.440

[15] Perry, A.B. (2004). Decreasing math anxiety in college students. *College Student Journal 38*(2), 321–324.

[16] Prosser, M., Martin, E., Trigwell, K., Ramsden, P. and Middleton, H. (2007). University academics' experience of research and its relationship to their experience of teaching. *Instructional Science 36*(1), 3–16. https://doi.org/10.1007/s11251-007-9019-4.

[17] Rappuoli, R., Pizza, M., Del Giudice, G. and De Gregorio, E. (2014). Vaccines, new opportunities for a new society. *Proceedings of The National Academy Of Sciences 111*(34), 12288–12293. https://doi.org/10.1073/pnas.1402981111

[18] Roberts, B., Caspi, A. and Moffitt, T. (2003). Work experiences and personality development in young adulthood. *Journal of Personality and Social Psychology, 84*(3), 582–593. https://doi.org/10.1037/0022-3514.84.3.582

[19] Russell, B. (1968). The impact of science on society. AMS Press.

[20] Russell, S. (2017). Artificial intelligence: The future is superintelligent. *Nature 548*(7669), 520–521. https://doi.org/10.1038/548520a

[21] Shams, S. and Thrassou, A. (2018). Theorization and industry-based research project development: Bridging the industry—academia research gap. *Industry and Higher Education 33*(2), 79–82. https://doi.org/10.1177/0950422218797919

[22] Stevens, B. (2005). What communication skills do employers want? Silicon valley recruiters respond. *Journal of Employment Counseling, 42*(1), 2-9. https://doi.org/10.1002/j.2161-1920.2005.tb00893.x

[23] Stott, T., Zaitseva, E. and Cui, V. (2012). Stepping back to move forward? Exploring Outdoor Education students' fresher and graduate identities and their impact on employment destinations. *Studies in Higher Education 39*(5), 711–733. https://doi.org/10.1080/03075079.2012.743116

[24] Tamir, P. and Pilar-Garcia, M. (1992). Characteristics of laboratory exercises included in science textbooks in Catalonia (Spain). *International Journal of Science Education, 14*(4), 381–392. https://doi.org/10.1080/0950069920140402

[25] Tom, G., Burns, M. and Zeng, Y. (1997). Your life on hold: the effect of telephone waiting time on customer perception. *Journal of Direct Marketing, 11*(3), 25-31. https://doi.org/10.1002/(sici)1522-7138(199722)11:3<25::aid-dir5>3.0.co;2-z

[26] Towns, M., Harwood, C., Robertshaw, M., Fish, J. and O'Shea, K. (2015). The Digital Pipetting Badge: A Method to Improve Student Hands-on Laboratory Skills. *Journal of Chemical Education 92*(12), 2038–2044. https://doi.org/10.1021/acs.jchemed.5b00464

[27] Zohuri, B. (2016). Electricity, an Essential Necessity in our Life. *Application of Compact Heat Exchangers for Combined Cycle Driven Efficiency in Next Generation Nuclear Power Plants*, 17–35. https://doi.org/10.1007/978-3-319-23537-0_2

Results on Basic Reproduction Number

Shilpa Umdekar[1] **and Praveen Kumar Shrama**[2]

ABSTRACT

The present paper attempts to highlight the behaviour of basic reproduction number, $\Re_0$ *with homogeneous transmission function and Holling type II treatment function. It further traces the evolution of* $\Re_0$ *in the context of various functions. It is a significant epidemiology concept, which is used to figure out whether the outbreak of disease is inevitable.*

Keywords: Basic Reproduction Number, Holling Type II Function, Homogeneous Transmission Function.

INTRODUCTION

Infectious diseases have been a part of human life since time immemorial. Evidence tells us that epidemics often end up causing mass deaths. Increase in healthcare has helped to put a check on infectious diseases to some extent. However, in recent years it has again emerged as a challenge. Diseases like HIV, acute encephalitis syndrome, and more recently COVID-19, caused high mortality. Whenever, we talk about epidemic or endemic, a very significant threshold quantity, reproduction number surfaces very often. These days, it's in trend to assess the disease pattern, outbreak and control on the basis of reproduction number. This measure is often used to estimate the probability of an epidemic or pandemic outbreak of developing infectious diseases. More precisely, the basic reproduction number, $\Re_0$ is the expected number of secondary infections a single infected individual can produce when introduced in population of susceptibles. When $\Re_0$ is higher than one, disease outbreak will occur; when $\Re_0$ is lower than one disease often fails to spread and may remain endemic in population. In order to evaluate true condition more accurately, it is necessary to measure whether value of $\Re_0$ is higher or lower than one.

---

[1] Research Scholar, Shri Vaishnav Institute of Science, Shri Vaishnav Vidyapeeth Vishwavidyalaya, Indore, Madhya Pradesh, India. E-mail: shilpaumdekar12@gmail.com

[2] Associate Professor, Shri Vaishnav Institute of Science, Shri Vaishnav Vidyapeeth Vishwavidyalaya, Indore, Madhya Pradesh, India. E-mail: praveen_jan1980@rediffmail.com

The concept of $\Re_0$ can be found in various systems, however, in this paper, the focus is on deterministic compartmental models using differential equations. The present work, studies some of the previous works done on the quantity and provides a model which gives a more plausible $\Re_0$ value to assess the disease breakout. The aim is to assess $\Re_0$ for an *SEIR* model with homogeneous transmission function and Holling type–II treatment function.

REVIEW OF LITERATURE

Wang & Ruan (2004) proposed and developed an SIR model using bilinear incidence rate and constant treatment function which describes the limited capacity of treatment by

$$h(I) = \begin{cases} a, I > 0 \\ 0, I = 0 \end{cases}$$

where $a > 0$ is a constant and responsible for treatment capacity of infectives. The model is beneficial to understand how disease dynamics may be influenced under limited resources of treatment. The reproduction number which depends upon infection rate β, recruitment rate A, natural death rate μ and natural recovery rate δ

$$\Re_0 = \frac{\beta A}{\mu(\mu + \delta)}$$

Note that the value of $\Re_0$ is free from $k(I)$ which is the removal rate of infected due to their treatment. To understand the effect of $k(I)$ the model is further analysed in respect of endemic equilibriums. Through further analysis they emphasize on the cost effective treatment capacity of a population and should be considered rather than large treatment capacity. The proposed treatment rate in this model is adequate for large population of infectives.

Wang (2006) proposed a modification of treatment function prior to the constant treatment function. He introduced and implemented a treatment function in which the treatment rate is directly proportional to the number of infections before the limit of treatment is reached, and when the maximum capacity is reached, the function takes its maximum saturated value.

$$h(I) = \begin{cases} aI, 0 < I < I_0 \\ aI_0, I > I_0 \end{cases}$$

He implemented the function in an SIR model with an additional parameter disease related death ϵ and concluded that bringing reproduction number down to below than one is not sufficient to eliminate a disease which in this model is given by

$$\Re_0 = \frac{\beta A}{\mu(\mu+\delta+\epsilon+a)}$$

Zhang & Xianning, (2008) introduced the saturated treatment function for the better analysis of real system through epidemic model. This function is widely known as Holling type–II treatment function.

$$h(I) = \frac{aI}{1+bI}$$

where $a > 0$ is a cure rate and $b \geq 0$ measures the magnitude of the consequence of the infected individual being delayed for treatment. The specialty of this function is that it is defined as continually differentiable and characterizes the situation of limited medical resources. They proposed this model with saturated incidence rate.

$$\Re_0 = \frac{\beta A}{\mu(\mu+\delta+\epsilon+a)}$$

In this model, $\Re_0$ doesn't remain the primary threshold quantity for evaluation of epidemic since value of $b > 0$ effects the behaviour of population which cannot be seen in the expression. The model is close to real system since it addresses the psychological effect of delayed care on infected population than prior ones. They insist on a timely and reliable treatment of infected to eradicate disease.

Pathak *et al.* (2010) proposed an SIR model with homogeneous transmission function with no permanent immunity. Transmission rate is defined using sociological and psychological parameters α_1 and α_2 respectively given by

$$\theta = \frac{\beta I}{1+\alpha_1 S+\alpha_2 I}$$

The introduction of these parameters by authors is a modified saturation effect. The reproduction number takes into account the sociological effect and is given by,

$$\Re_0 = \frac{\beta A - A\alpha_1(\mu+\delta)}{\mu(\mu+\delta)}$$

The reproduction number here does not depend upon α_2.This model indicates that reduction of social awareness in infected and rise in protective measure in susceptible are likely to have an effect on transmission of disease in the population.

PROPOSED MATHEMATICAL MODEL

The epidemic model proposed in the present work, is an SEIR model with saturated incidence rate and Holling type II treatment function using non-linear ordinary differential equations.

$$\frac{dS}{dt} = A - \frac{\beta SI}{1+\alpha_1 S+\alpha_2 I} - \mu S$$

$$\frac{dE}{dt} = \frac{\beta SI}{1+\alpha_1 S+\alpha_2 I} - (\gamma+\mu)E \qquad \dots (1)$$

$$\frac{dI}{dt} = \gamma E - (\delta+\mu)I - \frac{aI}{1+bI}$$

$$\frac{dR}{dt} = \delta I - \mu R + \frac{aI}{1+bI}$$

Where $S(t)$, $E(t)$, $I(t), R(t)$ denote the susceptible, exposed, infected and recovered number of individuals at time t. Other parameters used in the system are infection rate β, recruitment rate A, natural death rate μ, natural recovery rate δ, progression rate γ, sociological parameters $\alpha_1 > 0$, and psychological parameter $\alpha_2 > 0$, cure rate $a > 0$, magnitude of the consequence of the infected individual being delayed for treatment by $b \geq 0$.

Since first three equations are free from R, we can reduce the system (1) into,

$$\frac{dS}{dt} = A - \frac{\beta SI}{1+\alpha_1 S+\alpha_2 I} - \mu S$$

$$\frac{dE}{dt} = \frac{\beta SI}{1+\alpha_1 S+\alpha_2 I} - (\gamma+\mu)E \qquad \dots (2)$$

$$\frac{dI}{dt} = \gamma E - (\delta+\mu)I - \frac{aI}{1+bI}$$

REPRODUCTION NUMBER FOR PROPOSED MODEL

The next generation method was introduced by Diekmann *et al.*, (1990) and has been elaborated by van den Driessche & Watmough, (2002) to derive $\Re_0$ in the cases where two or more than two ODE compartments. The next generation method outlined by van den Driessche & Watmough, (2002) for the proposed model has been used. The system (2) will always have a disease-free equilibrium at $\left(\frac{A}{\mu}, 0, 0\right)$. For calculating $\Re_0$, the focus has been on infected compartments E and I only, thus, system furthermore reduces to,

$$\frac{dE}{dt} = \frac{\beta SI}{1+\alpha_1 S+\alpha_2 I} - (\gamma+\mu)E$$

$$\frac{dI}{dt} = \gamma E - (\delta+\mu)I - \frac{aI}{1+bI}$$

Let $x = (E, I)^T$. The system (3) can be written as $\frac{dx}{dt} = (F - V)x$ where

$$F = \begin{bmatrix} \frac{\beta SI}{1+\alpha_1 S+\alpha_2 I} \\ 0 \end{bmatrix} \text{ and } V = \begin{bmatrix} (\gamma+\mu)E \\ -\gamma E + (\delta+\mu)I + \frac{aI}{1+bI} \end{bmatrix}$$

The Jacobian matrix of F and V at disease free equilibrium $\left(\frac{A}{\mu}, 0, 0\right)$ are

$$F_1 = \begin{bmatrix} 0 & \frac{\beta A}{\mu+\alpha_1 A} \\ 0 & 0 \end{bmatrix}, V_1 = \begin{bmatrix} (\gamma+\mu) & 0 \\ -\gamma & (\delta+\mu)+a \end{bmatrix}$$

$$V_1^{-1} = \frac{1}{(\gamma+\mu)(\delta+\mu+a)}\begin{bmatrix}(\delta+\mu+a) & 0 \\ \gamma & (\gamma+\mu)\end{bmatrix}$$

The Next Generation Matrix is given by,

$$F_1V_1^{-1} = \begin{bmatrix}\frac{\gamma\beta A}{(\mu+\alpha_1 A)(\gamma+\mu)(\delta+\mu+a)} & \frac{\beta A}{(\mu+\alpha_1 A)(\delta+\mu+a)} \\ 0 & 0\end{bmatrix}$$

Reproduction number $\Re_0$ is the dominant eigen value of $F_1V_1^{-1}$,

$$\Re_0 = \frac{\gamma\beta A}{(\mu+\alpha_1 A)(\gamma+\mu)(\delta+\mu+a)} \qquad \dots (4)$$

DISCUSSION

With the aid of the expression derived in (4), $\Re_0$ value can be measured to determine the probability of epidemic. The presented model supports disease free equilibrium and endemic equilibrium. $\Re_0$ can further be used to study the stability of equilibrium for the proposed model. The reproduction number relies on social behaviour, not on psychological impact of infected in population. This indicates that reduction of social awareness in infected and rise in protective measure in susceptibles are likely to have an effect on transmission of disease in the population. In order to grasp the psychological effect of the model, some further estimates are required. In this model, $\Re_0$ does not depend on magnitude of the consequence of the infected individual being delayed for treatment, therefore, $\Re_0$ can no longer be used as the primary threshold quantity for disease assessments. There is a need to do some further assessment to understand the impact of b. The model is close to real system since it addresses the psychological effect of delayed care on infected population than prior ones. As discussed above, conclusions cannot be drawn solely on $\Re_0$ value, for a disease spread, further assessments are required before for making control strategies.

REFERENCES

[1] Diekmann, O., Heesterbeek, J. and Metz, J. (1990). On the definition and the computation of the basic reproduction ratio $\Re_0$ in models for infectious diseases. *J. Math. Biol.*, 35, 503–522.

[2] Pathak, S., Maiti, A. and Samanta, G. (2010). Rich dynamics of an SIR epidemic model. *Nonlinear Analysis: Modelling and Control*, 15, 71–81.

[3] Van den Driessche, P. and Watmough, J. (2002). Reproduction numbers and sub-threshold endemic equilibria for compartmental models of disease transmission. *Math. Biosci.*, 180, 29–48.

[4] Wang, W. (2006). Backward bifurcation of an epidemic model with treatment. *Mathematical Biosciences*, 201, 58–71.

[5] Wang, W. and Ruan, S. (2004). Bifurcation in an epidemic model with constant removal rate of infecticves. *Journal of Mathematical Analysis and Applications*, 291(2), 775–793.

[6] Zhang, X. and Xianning, L. (2008). Backward bifurcation of an epidemic model with saturated treatment function. *Journal of Mathematical Analysis and Applications*, 348, 433–443.

Review of Different Deep Learning Models to Predict Human Mental State

Shubham Kothari[1] **and Anand Singh Rajawat**[2]

ABSTRACT

It is found that psychological instability is one of the most widely recognized illnesses like diabetes, heart diseases etc. Depression is mental disorder that has affected people all over the world. It can lead to suicidal attempts at the worst stage. In order to find out mental state of individuals, different deep learning models have been used. In the present paper different techniques of deep learning have been studied and focuses on image analysis and classification to predict depressed individuals and propose a multi-model to identify depressed individuals.

Keywords: CNN, Deep Learning, MDD, Mental State Assessment, RNN.

INTRODUCTION

Now-a-days depression is also referred as Major Depressive Disorder (MDD) (Belmaker & Agam, 2008). An individual might be experiencing depression if he or she give indications of extreme irritability, outrage, loss of enthusiasm for most loved activities, constant lament, thoughts of death or self-destruction, aches or pain, migraine, issues of stomach related problems without any physical cause over a period of time. The reasons of depression can be personal or family record of depression, significant life changes, injury or stress, certain physical diseases and drugs (Pampouchidou *et al.*, 2017).

The essential point is early recognition of users showing indications of depression so that accurate diagnosis and treatment may start at an early stage. Depression if untreated may prompt unwanted effects apart from the disease itself. For example, increased chances of dangerous conduct like drug or liquor addiction make it hard to recover from different diseases and the disease itself may be prolonged for considerable span of time. The numbers are very upsetting as

[1] Research Scholar, Shri Vaishnav Institute of Information Technology, Shri Vaishnav Vidyapeeth Vishwavidyalaya, Indore, Madhya Pradesh, India. E-mail: shubhever.cs@gmail.com

[2] Assistant Professor, Shri Vaishnav Institute of Information Technology, Shri Vaishnav Vidyapeeth Vishwavidyalaya, Indore, Madhya Pradesh, India. E-mail: anandrajawatds@gmail.com

there are roughly 300 million individuals experiencing depression issue worldwide and this number is expected to grow each year (Valstar *et al.*, 2013). It has been found that the majority of individuals experiencing depression is from low-income families (Spitzer *et al.*, 2000). Clearly there is a solid need to handle depression by identifying risks as quickly as possible. Recognizing depressed individuals can help to find and help needy people who need special care.

Deep belief networks can be used for Image Recognition, Video Recognition and Motion Capture Data (Hua *et al.*, 2015). It uses probabilities and unsupervised learning. They are composed of binary latent variables and contain both types of layers i.e., directed and undirected. Identifying depression from images is framed as a classification problem in the field of Deep Learning (DL). Identifying symptoms from image can give critical information about a user regarding his or her depression diagnosis. The current methodology using supervised machine learning to classify a tweet having depressive symptoms is based on exact string matching feature (Spitzer *et al.*, 2000). This method falls short on classifying different users' post having different intentions.

REVIEW OF LITERATURE

In the field of image analysis and classification, numerous researchers have made appreciable efforts to create the several methods of image classification. Various efforts have been undertaken to address the analysis of image datasets. The comparison table of different methods involved in the depression detection are shown (Table 1).

Table 1: Comparison of Different Deep Learning Methods

S. No.	*Paper*	*Author*	*Year*	*Dataset*	*Data features*	*Results*
1.	AVEC 2014 – 3D Dimensional Affect and Depression Recognition Challenge	Michel *et al.*	2014	AVEC 2013 audio-visual depression corpus	Audio sets and textual features from manual transcripts	Detection of the active dimensions arousal, valence and dominance in continuous time.
2.	Inferring Clinical Depression From Speech And Spoken Utterances	Asgari *et al.*	2014	INTERSPEECH 2010 Paralinguistic Challenge [18] using openSMILE toolkit	Audio sets and textual features from manual transcripts	Problem of detecting depression from recordings spoken utterances can be used.
3.	Artificial Intelligent System for Automatic Depression Level Analysis through Visual and Vocal Expressions	Jain *et al.*	2017	AVEC2014 dataset	Audio sets and textual features from manual transcripts	Artificial intelligent system was used for automatic depression scale prediction.

S. No.	*Paper*	*Author*	*Year*	*Dataset*	*Data features*	*Results*
4.	Voxel wise 3D Convolutional and Recurrent Neural Networks for Epilepsy and Depression Diagnostics from Structural and Functional MRI Data	Pominova *et al.*	2018	T2-weighted MRI EPI series and T1-weighted MP-RAGE images	Temporal MRI images during depression diagnostics	I. VoxCNN-B model achieves the highest results in EvsNE and EvsH tasks
5.	Deep learning for online harassment detection in Tweets	Marwa *et al.*	2018	Twitter dataset	A huge chunk of tweets used as strings	Deep learning models namely CNN, LSTM and BLSTM outperform the other classification methods
6.	Detecting Depression Severity by Interpretable Representations of Motion Dynamics	Kacem *et al.*	2018	Dataset based on clinical trial for treatment of depression.	Set of questions based on experiments carried on depression	Method to measure depression severity from facial and head movement dynamics.
7.	Sentiment Analysis Based on Deep Learning and its aapplication in Screening for Perinatal Depression [8]	Chen *et al.*	2018	Data of WeChat friends circle	Textual data	Deep learning LSTM network model can be used for perinatal depression screening to achieve results
8.	Human Behaviour-based Automatic Depression Analysis using Hand-crafted Statistics and Deep Learned Spectral Features	Song *et al.*	2018	AVEC 2016 DAIC-WOZ database	Clinical Interviews of clinical patients in form of audios and videos	Only AUs, gaze directions and head poses have been used as the human behaviour signals for the depression analysis
9.	Depression Severity Classification from Speech Emotion	Harati *et al.*	2018	Videos of subjects were collected as part of an ongoing DBS	Data in video form	We study the proposed approach experimentally and compare its effectiveness in classifying depression severity.
10.	Video-Based Depression Level Analysis by Encoding Deep Spatiotemporal Features	Jazaery *et al.*	2018	AVEC2014 Dataset and AVEC2013 Depression Dataset	Audio sets and textual features from manual transcripts	Both AVEC2013 and AVEC2014 datasets have shown that our visual-based approach is promising
11.	Context-aware Deep Learning For Multi-modal Depression Detection	Lam *et al.*	2019	DAIC-WOZ dataset	Clinical interviews of clinical patients in form of audios and videos	Deep 1D CNN and Transformer models achieved state-of the-art performance for audio and text modalities

S. No.	Paper	Author	Year	Dataset	Data features	Results
12.	A Knowledge-Based Recommendation System that includes Sentiment Analysis and Deep Learning	Rosa *et al.*	2019	Named Entity Recognition (NER) datasets	Data in text form.	Performance assessment of eSM and eSM2 metrics was performed, and results obtained by the eSM2 were superior in the perceptual evaluation of the RS.
13.	Fusion of deep- learned and hand crafted features for cancelable recognition systems.	Abdellatef *et al.*	2020	Images dataset name not mentioned	Full body images with face and palm	Deep learned features and hand crafted features are extracted to identify face, iris, palm and finger print.

METHOD

Numerous researches have been performed which focused on studying psychological health issues. Some questionnaire methods are also used to identify symptoms of the person. These clinical methods may fail as they are dependent on the memory of the user and the questionnaire variations. Some recent research focuses on the text, audio and video only rather than facial expressions. But when we add both of these factors i.e. depression and image analysis of linguistic parameters; only a few researches came to our knowledge which identifies depressed individuals using deep learning techniques.

The present paper focuses on developing a deep neural network architecture which will first detect face in the given image using VGG-Face (CNN) and Shifted Filter Response (deep belief network) and then identify symptoms of depression on that face. Deep belief networks are subset of deep neural networks algorithms that are based on human brain, having ability to recognize patterns and process complex information (Hua *et al.*, 2015). These opportunities inspire to propose multimodel on the image analysis for depression symptoms using VGG approach in Convolutional Neural Network and Shifted Filter Response in Deep Belief Network.

Multimodel will have multiple convolutional layers, multiple Rectified Linear Units (ReLu), multiple (max pooling) layers and fully connected layers. It is combination of VGG approach in CNN and Shifted Filter response in deep belief network. VGG Face is an extended CNN used for acquisition of image. It is used for performance improvement of existing CNN (Table 2). The proposed model deals with depression recognition from single image or a set of images tracked in a video. DBN stands for Deep Belief Networks. COSFIRE (Combination of Shifted FIlter REsponses) is a trainable filter method which can be used for pattern recognition in numerous computer vision tasks.

Table 2: Datasets for Depression Detection Systems

Dataset	*Modality*	*Depression Annotation*
AVEC 2013 (Zhu *et al.*, 2017)	A + V V	BDI-II (300 videos) (3ideos)
AVEC 2014 (Valstar *et al.*, 2014) (Jan *et al.*, 2017)	A + V	BDI-II
AVEC 2017 (L. Yang *et al.*, 2017)	A + V	BDI-II
Crisis Text Line	T	Manual annotation for depression
Dementia Bank Database	A + T + V	PHQ 8 (65,024 forum posts)
DAIC	A + T + V	HAMD

The methodology for the proposed work is as follows:

- Data Collection—In the data collection phase, datasets are derived from image datasets (Al Jazaery & Guo, 2018). These image datasets contains collection of different images of millions of users who are struggling with depression. These images can be utilized for learning the identification of depressed person.
- Image Classification—After data collection, image classification will be done whether the image contains the individual's face or not.
- Feature Extraction—To extract features from image, existing deep learning neural networks such as CNN is used.
- Designing Multimodel—To extract the features of depression from an image, a multimodel architecture is designed and gets it trained by millions of images of different users provided by trained datasets.
- Image Categorization—Supervised classifiers are created by learning the types of images from existing images in the datasets. The learned classifier will be used for classification of depressed users (Figure 1).

Image Dataset are used to train proposed model (Pampouchidou *et al.*, 2017). Scala and Python are used for larger data set processing. Spark tool can be used on Hadoop for the distributed storage and processing of big data sets. Hadoop operates by splitting files into large blocks of data and then distributing those datasets across the nodes in a cluster. It then transfers code into the nodes, for processing data in parallel. IBM Watson tool provides environment for big data processing. It helps data scientists and developers accelerate AI and machine-learning deployment. Huge amount of dataset can be uploaded on Google Drive on paid basis or Amazon Web Services (AWS) which provides on demand cloud computing platforms and APIs.

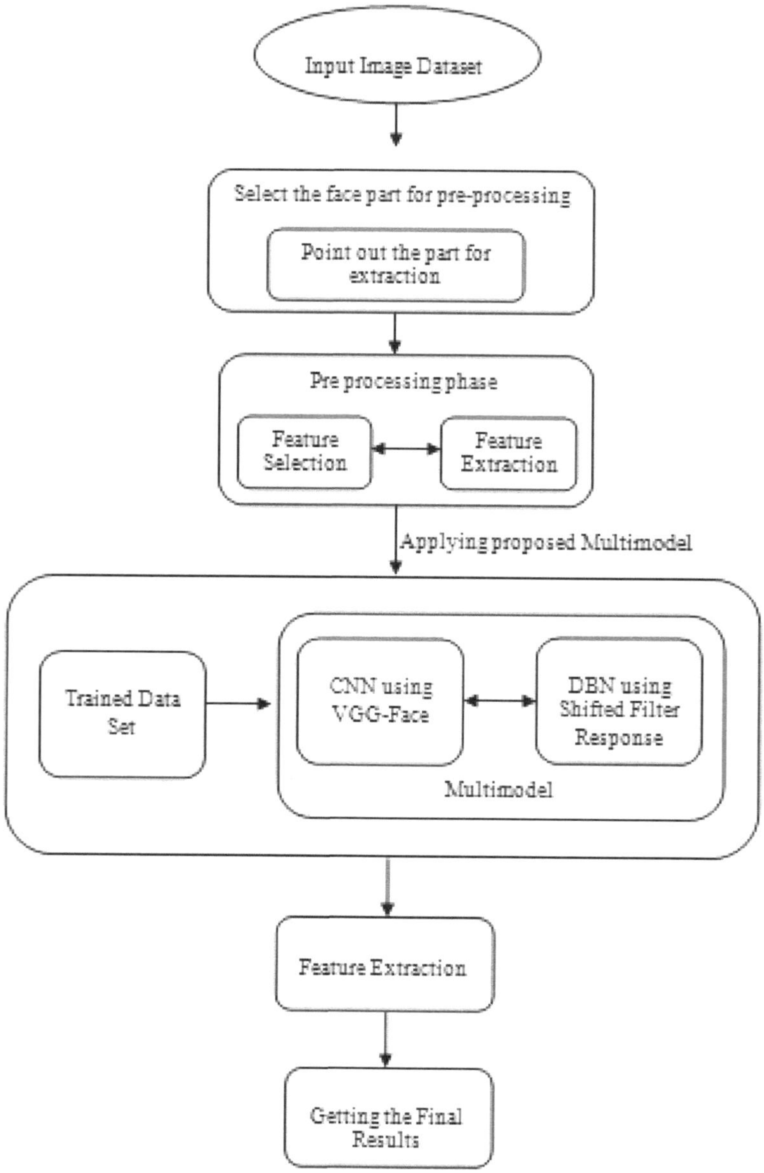

Fig. 1: Flow Diagram of Proposed Multimodel

CONCLUSION

Prediction accuracy can be improved by focusing on both image analysis of facial expression and motion dynamics of eyes, lips, head and hand movement to compute time and space complexity. Deep learning algorithms generally process complex and large datasets containing a large number of features to get meaningful information. In most cases, these algorithms have high computational complexity issues while processing large datasets due to the presence of lesser weight features (irrelevant or redundant). As a result, the computing increases with increase in the number of features, number of records, feature types and inbuilt feature categories present in datasets. Appropriate feature selection before applying Deep Learning Algorithms are used as a solution to resolve the computing speed and accuracy issues effectively during processing of large datasets.

REFERENCES

[1] Abdellatef, E., Omran, E.M., Soliman, R.F., Ismail, N.A., Abd Elrahman, S.E.S.E., Ismail, K.N., Rihan, M., Abd El-Samie, F.E. and Eisa, A.A. (2020). Fusion of deep-learned and hand-crafted features for cancelable recognition systems. *Soft Computing, 24*(20), 15189–15208.

[2] Al Jazaery, M. and Guo, G. (2018). Video-based depression level analysis by encoding deep spatiotemporal features. *IEEE Transactions on Affective Computing.*

[3] Andalibi, N., Ozturk, P. and Forte, A. (2015). Depression-related imagery on instagram. *Proceedings of the 18th ACM Conference Companion on Computer Supported Cooperative Work & Social Computing*, 231–234.

[4] Asgari, M., Shafran, I. and Sheeber, L.B. (2014). Inferring clinical depression from speech and spoken utterances. *2014 IEEE International Workshop on Machine Learning for Signal Processing (MLSP)*, 1–5.

[5] Badrinarayanan, V., Kendall, A. and Cipolla, R. (2017). Segnet: A deep convolutional encoder-decoder architecture for image segmentation. *IEEE Transactions on Pattern Analysis and Machine Intelligence, 39*(12), 2481–2495.

[6] Belmaker, R.H. and Agam, G. (2008). Major depressive disorder. *New England Journal of Medicine, 358*(1), 55–68.

[7] Bosley, J.C., Zhao, N.W., Hill, S., Shofer, F.S., Asch, D.A., Becker, L.B. and Merchant, R.M. (2013). Decoding twitter: Surveillance and trends for cardiac arrest and resuscitation communication. *Resuscitation, 84*(2), 206–212.

[8] Chen, Y., Zhou, B., Zhang, W., Gong, W. and Sun, G. (2018). Sentiment analysis based on deep learning and its application in screening for perinatal depression. *2018 IEEE Third International Conference on Data Science in Cyberspace (DSC)*, 451–456.

[9] Cohn, J.F., Cummins, N., Epps, J., Goecke, R., Joshi, J. and Scherer, S. (2018). Multimodal assessment of depression from behavioral signals. In *The Handbook of Multimodal-Multisensor Interfaces: Signal Processing, Architectures, and Detection of Emotion and Cognition,* Vol. 2, pp. 375–417.

[10] De Choudhury, M., Counts, S., Horvitz, E.J. and Hoff, A. (2014). Characterizing and predicting postpartum depression from shared facebook data. *Proceedings of the 17th ACM Conference on Computer Supported Cooperative Work & Social Computing*, 626–638.

[11] De Choudhury, M., Gamon, M., Counts, S. and Horvitz, E. (2013). Predicting depression via social media. *Proceedings of the International AAAI Conference on Web and Social Media, 7*(1).

[12] De Choudhury, M., Kiciman, E., Dredze, M., Coppersmith, G. and Kumar, M. (2016). Discovering shifts to suicidal ideation from mental health content in social media. *Proceedings of the 2016 CHI Conference on Human Factors in Computing Systems*, 2098–2110.

[13] Harati, S., Crowell, A., Mayberg, H. and Nemati, S. (2018). Depression severity classification from speech emotion. *2018, 40th Annual International Conference of the IEEE Engineering in Medicine and Biology Society (EMBC)*, 5763–5766.

[14] He, L. and Cao, C. (2018). Automated depression analysis using convolutional neural networks from speech. *Journal of Biomedical Informatics, 83*, 103–111.

[15] Hua, Y., Guo, J. and Zhao, H. (2015). Deep belief networks and deep learning. *Proceedings of 2015 International Conference on Intelligent Computing and Internet of Things*, 1–4.

[16] Jan, A., Meng, H., Gaus, Y.F.B.A. and Zhang, F. (2017). Artificial intelligent system for automatic depression level analysis through visual and vocal expressions. *IEEE Transactions on Cognitive and Developmental Systems, 10*(3), 668–680.

[17] Kacem, A., Hammal, Z., Daoudi, M. and Cohn, J. (2018). Detecting depression severity by interpretable representations of motion dynamics. *2018 13th IEEE International Conference on Automatic Face & Gesture Recognition (FG 2018)*, 739–745.

[18] Kharel, P., Sharma, K., Dhimal, S. and Sharma, S. (2019). Early detection of depression and treatment response prediction using machine learning: a review. *2019 Second International Conference on Advanced Computational and Communication Paradigms (ICACCP)*, 1–7.

[19] Lam, G., Dongyan, H. and Lin, W. (2019). Context-aware deep learning for multi-modal depression detection. *ICASSP 2019-2019 IEEE International Conference on Acoustics, Speech and Signal Processing (ICASSP)*, 3946–3950.

[20] Leng, B., Zhang, X., Yao, M. and Xiong, Z. (2014). 3d object classification using deep belief networks. *International Conference on Multimedia Modeling*, 128–139.

[21] Liu, S. and Deng, W. (2015). Very deep convolutional neural network based image classification using small training sample size. *2015 3rd IAPR Asian Conference on Pattern Recognition (ACPR)*, 730–734.

[22] Marwa, T., Salima, O. and Souham, M. (2018). Deep learning for online harassment detection in tweets. *2018 3rd International Conference on Pattern Analysis and Intelligent Systems (PAIS)*, 1–5.

[23] Orabi, A.H., Buddhitha, P., Orabi, M.H. and Inkpen, D. (n.d.). *Deep Learning for Depression Detection of Twitter Users.*

[24] Pampouchidou, A., Pediaditis, M., Maridaki, A., Awais, M., Vazakopoulou, C.-M., Sfakianakis, S., Tsiknakis, M., Simos, P., Marias, K. and Yang, F. (2017). Quantitative comparison of motion history image variants for video-based

depression assessment. *EURASIP Journal on Image and Video Processing, 2017*(1), 1–11.

[25] Pampouchidou, A., Simos, P.G., Marias, K., Meriaudeau, F., Yang, F., Pediaditis, M. and Tsiknakis, M. (2017). Automatic assessment of depression based on visual cues: A systematic review. *IEEE Transactions on Affective Computing, 10*(4), 445–470.

[26] Pominova, M., Artemov, A., Sharaev, M., Kondrateva, E., Bernstein, A. and Burnaev, E. (2018). Voxelwise 3d convolutional and recurrent neural networks for epilepsy and depression diagnostics from structural and functional mri data. *2018 IEEE International Conference on Data Mining Workshops (ICDMW)*, 299–307.

[27] Qiu, S., Chang, G.H., Panagia, M., Gopal, D.M., Au, R. and Kolachalama, V.B. (2018). Fusion of deep learning models of MRI scans, Mini–Mental State Examination, and logical memory test enhances diagnosis of mild cognitive impairment. *Alzheimer's & Dementia: Diagnosis, Assessment & Disease Monitoring, 10*, 737–749.

[28] Ragab, D.A., Sharkas, M., Marshall, S. and Ren, J. (2019). Breast cancer detection using deep convolutional neural networks and support vector machines. *PeerJ, 7*, e6201.

[29] Rosa, R.L., Schwartz, G.M., Ruggiero, W.V. and Rodríguez, D.Z. (2018). A knowledge-based recommendation system that includes sentiment analysis and deep learning. *IEEE Transactions on Industrial Informatics, 15*(4), 2124–2135.

[30] S'adan, M.A.H.M., Pampouchidou, A. and Meriaudeau, F. (2018). Deep learning techniques for depression assessment. *2018 International Conference on Intelligent and Advanced System (ICIAS)*, 1–5.

[31] Song, S., Shen, L. and Valstar, M. (2018). Human behaviour-based automatic depression analysis using hand-crafted statistics and deep learned spectral features. *2018 13th IEEE International Conference on Automatic Face & Gesture Recognition (FG 2018)*, 158–165.

[32] Spitzer, R.L., Williams, J.B.W., Kroenke, K., Hornyak, R., McMurray, J. and Group, P.H.Q.O.-G. S. (2000). Validity and utility of the PRIME-MD patient health questionnaire in assessment of 3000 obstetric-gynecologic patients: the PRIME-MD Patient Health Questionnaire Obstetrics-Gynecology Study. *American Journal of Obstetrics and Gynecology, 183*(3), 759–769.

[33] Valstar, M., Schuller, B., Smith, K., Almaev, T., Eyben, F., Krajewski, J., Cowie, R. and Pantic, M. (2014). Avec 2014: 3d dimensional affect and depression recognition challenge. *Proceedings of the 4th International Workshop on Audio/Visual Emotion Challenge*, 3–10.

[34] Valstar, M., Schuller, B., Smith, K., Eyben, F., Jiang, B., Bilakhia, S., Schnieder, S., Cowie, R. and Pantic, M. (2013). Avec 2013: the continuous audio/visual emotion and depression recognition challenge. *Proceedings of the 3rd ACM International Workshop on Audio/Visual Emotion Challenge*, 3–10.

[35] Yang, F., Yang, J., Jin, Z. and Wang, H. (2018). A fusion model for road detection based on deep learning and fully connected CRF. *2018 13th Annual Conference on System of Systems Engineering (SoSE)*, 29–36.

[36] Yang, L., Jiang, D., Xia, X., Pei, E., Oveneke, M.C. and Sahli, H. (2017). Multimodal measurement of depression using deep learning models. *Proceedings of the 7th Annual Workshop on Audio/Visual Emotion Challenge*, 53–59.

[37] Zhu, Y., Shang, Y., Shao, Z. and Guo, G. (2017). Automated depression diagnosis based on deep networks to encode facial appearance and dynamics. *IEEE Transactions on Affective Computing*, *9*(4), 578–584.

The Information Needs and Seeking Behaviour of Faculty Members of a Private University

A Self Study

Pradyumna Kumar Panigrahi[1] **and Ravindra Kumar**[2]

ABSTRACT

This study examines the information needs and seeking behaviour of faculty members of Sai-Tirupati University, Udaipur. A structured questionnaire related to the study was distributed among 87 faculty members, out of which 62 were returned with a response of 71.2%. The data obtained through questionnaires were analyzed. Survey results indicate that respondents use the internet as their main source and channel for seeking information and e-books/e-journals are the major information resources used.

Keywords: Faculty, Information, Information Seeking Behaviour, Internet.

INTRODUCTION

Information is indispensable for any library as it is required by users. Growing demand of access by library users the quantity of information required in the library is huge. Users need information in every aspect of their lives to flourish. The library staff has to recognize the underlying need of the information seekers so that they can facilitate them in locating them in the libraries (Fiancor and Adams, 2004). According to Hoshvosky and Marsey (1968), "Information is a process which occurs within the human mind, when a problem and data for its solutions are brought into a productive union". The identification of information needs is a prerequisite for the design and development of any information system for providing information services based on the needs of the customers. This is a topic of great concern and interest to the information professionals as they seek, extract and prefer to use information.

The user of information is happy when he/she gets the required information. However, this is not always the case. Informational professional's responsibility

---

[1] Research Scholar, Department of Library & Information Science, Mandsaur University, Mandsaur, Madhya Pradesh, India. E-mail: pkpanigrahi677@gmail.com

[2] Assistant Professor, Department of Library & Information Science, Mandsaur University, Mandsaur, Madhya Pradesh, India. E-mail: ravindra.kumar@meu.edu.in

is to ensure maximum satisfaction of the user concerned and to minimize the time and difficulty in searching the required information. Surveys as a whole are well established in forecasting, and according to Wilson (1980), it would be important to know all the texts that an interviewer can use before an optimal collection. Creating an information system based on the needs of the customer community should be expected or considered. According to freedictionary.com, a faculty is the teacher or an instructor of a school or college, or of one of its divisions, especially those who are full-time employees. They provide academic guidance to students and expands the horizons of his/her knowledge through research and publications. This makes information mandatory for teachers.

INFORMATION SEEKING BEHAVIOUR

Information seeking behaviour (ISB) is a psychological happening. Library users seek information from various sources such as books, journals, magazines, newspapers, e-books, e-journals, research reports, audio-visuals, etc. Therefore, the library requires extensive library material and knowledgeable library staff, who can take care of the needs and behaviours of its users. Studying on ISB will be helpful for the library, because if the library knows about the exact needs and seeking behaviour of its users, it will help to improve the quality resources of library services. The present era is the era of the information and knowledge revolution. The growth in the information accessible on the Internet has significantly impacted information-seeking behaviour (Natrajan, 2012).

Information-Seeking behaviour derives from a supposed desire for information by the user in Wilson's (1981, 1997) models. In addition to this request, the user may have query about the information system (e.g. library or database) and other data sources (e.g. textbook, lecturer, handouts and suchlike). According to Krikelas (1983), information-seeking behaviour refers to any human action that is carried out to find a message that fits perceived needs. Wilson (2000) has described that information-seeking behaviour is designed to seek information and it is a function of the desire to fulfil a particular objective.

STUDY AREA

The researcher selected the faculty members of Sai-Tirupati University located in Udaipur, Rajasthan for the study. This is a private university and it was established in 2016. This university has been well advancing to reach heights in the sphere of global health research like-pharmacy, para-medical sciences, medicine, dental, and so on. The library of this university is well stocked with the latest books, journals, magazines, e-books, e-journals, etc.

STATEMENT OF THE PROBLEM

Faculty members are usually busy with teaching, preparing lessons, and attending conferences/seminars. As it is a medical university, faculty members are involved in teaching as well as clinical work. Therefore, they have to remain updated with the current developments in their specific subject area. Overflow and availability of bulk information in different formats creates information clutter which is time consuming and hinders the efforts of the faculty to get the relevant information. To meet the specific needs of the users/faculty, the librarian must find ways to facilitate them. In such a situation, user survey is needed to explore the strengths and weaknesses of the existing resources, to increase the efficiency and effectiveness of services. This calls for the introduction of new systems in the library to meet the diverse needs of library users. The present study is an attempt to assess the information needs and detect the seeking behavior of the faculty.

OBJECTIVES OF THE STUDY

The objectives of the present study are:

- To examine the different kinds of sources and channels followed when the faculty members need information.
- To determine the purpose of information seeking by the faculty members.
- To determine the various library resources used by the faculty members.
- To examine the library services and their use by the faculty members.
- To study the problem faced by the faculty members while seeking information from the library.

REVIEW OF LITERATURE

Numerous studies have been conducted on information needs and seeking behaviour. Muhammad and Kanwal (2009) identified that faculty members of the university use both digital and print data to satisfy their favourite requirements for knowledge. The respondents stated that the books were the primary source of knowledge as compared to journals. Patitungkho and Deshpande (2005) in their research at the Rajabhat University, Bangkok, Thailand found that 57% of the faculty members used textbooks, 82% sought lecture preparation information, 74% accessed Thai-language information, 24% used it in English, and 2% used French and Japanese. Suriya, *et al.* (2004) talk about the information-seeking behaviour of teaching faculty related to government art colleges in Cuddalore, Tamil Nadu. The results indicated that the teaching faculty visited the library to meet their educational needs. Curtis *et al.* (1997) conducted a study on the information needs of health science teaching staff. This study revealed that medical teaching faculty used e-resources to gather information. This analysis was performed among the teaching faculty of all disciplines of pharmacy, medicine in nursing at the University of Illinois, Chicago.

METHOD

The data were obtained from the faculty members of Sai-Technical University, Udaipur through survey/email/online survey methods. The structured questionnaire was circulated to eighty-seven faculty members, containing both open-ended and closed-ended questions. Members of the faculty were asked to fill out the survey. The answers were received from sixty-two faculty members and these were tabulated and further statistical analysis was carried out.

DATA ANALYSIS AND FINDINGS

A total of 62 responses were received, with the percentage being 71.26%. The analyses of the collected data are presented in Table 1 as in tabular form and in Figure 1 as in graphical form as per the percentage of respondents.

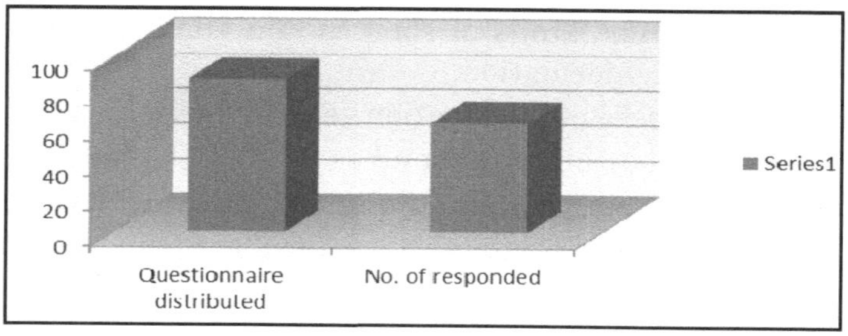

Fig. 1: Distribution of Questionnaire and Response Received

Sources and Channels of Information Seeking

The Figure 2 explains the uses of information sources and channels used by the respondents of the study. The data analysis reveals that a majority of the faculty

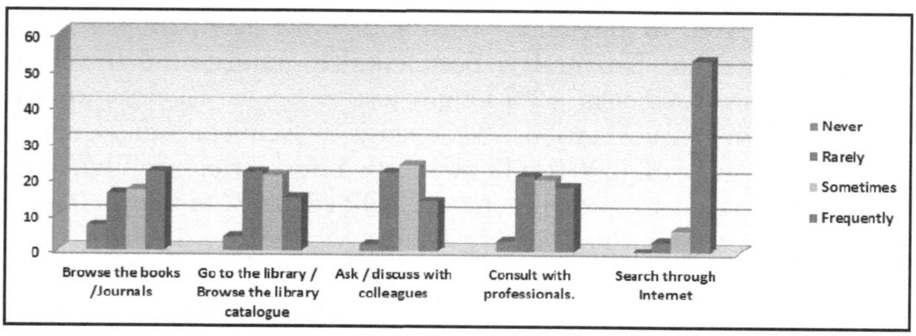

Fig. 2: Sources and Channels of Information Seeking by the Respondents

members i.e. 53 (85.8%) were using the Internet as major sources and channels for searching the Information frequently, 22 (35.48%) of the respondents were browsing the books/journals frequently, whereas 18 (29.03%) of the faculty members consulted professionals frequently.

Purpose of Information Seeking

The data analysis for information seeking is available in Figure 3 and reveals that 52 (83.87%) of the respondents were seeking information from the library for teaching/guiding students, 48 (77.42%) of the respondents were seeking information to keep up with current developments and 40 (64.52%) of respondents were seeking information for guiding research scholars.

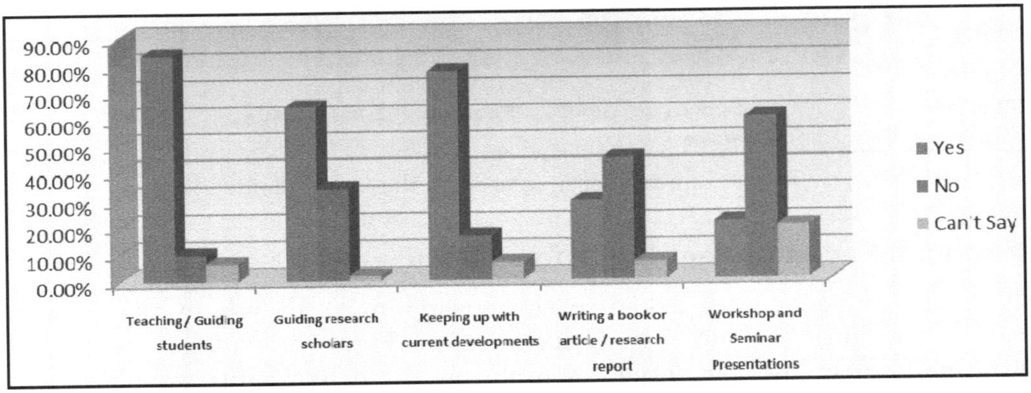

Fig. 3: Purpose of Information Seeking

Library Resources Used by the Respondents

The data analysis of Library resources used by the respondents is available in the Figure 4 and it is found in the table that 58 (93.55%) of the respondents use E-books/E-Journals, 55(88.71%) of the respondents used journals/magazines and 54 (87.10%) of the respondents used digital library.

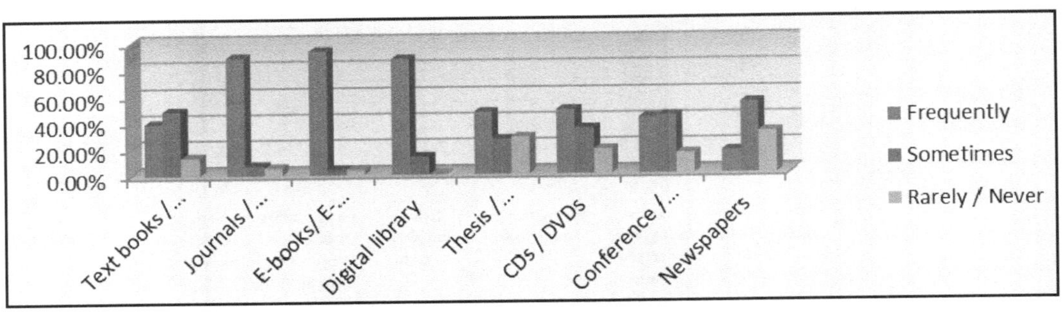

Fig. 4: Library Resources Used by the Respondents

Library Services Used by the Respondents

It is observed in Figure 5 the data analysis of the library services used by the respondents of the study. The data reveals that 54 (87.10%) of the respondents were using the circulation services frequently, 49 (79.03%) of the respondents were using the reference services, and 50 (80.65%) of the respondents were using the OPAC services.

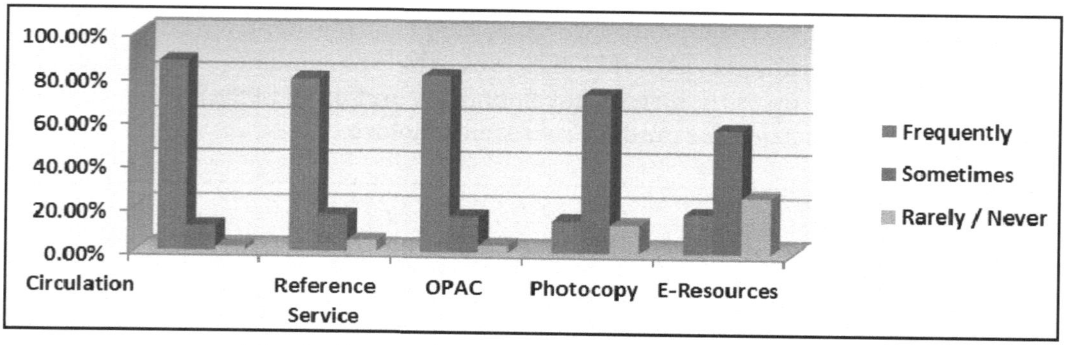

Fig. 5: Library Services used by the Respondents

Problems Faced in Seeking Information

All the respondents were asked in the questionnaire to submit the problems faced in the library while seeking information. Figure 6 explains displays the data analyzed in the percentage of problems faced by faculty members while obtaining information in the library. The respondent of 1 (1.61%) faculty member is facing the problems of unavailability of required/latest materials in the library. The library timing is not suitable for 2(3.23%) respondents. Lack of technical support is one of the problems for 5 (8.26%) of respondents. Similarly, lack of training in using the electronic resources is a problem for 7(11.29%) of respondents in their seeking of information, and 55 (88.71%) of the respondents are facing the problems of too many classes or administrative function in their scheduled work.

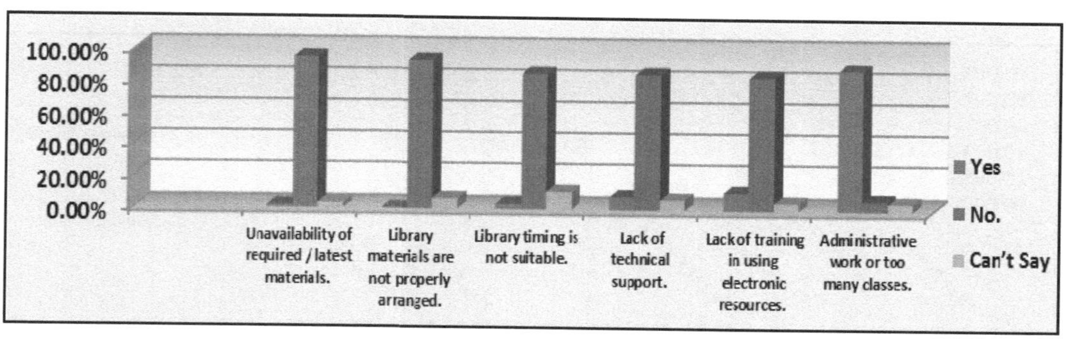

Fig. 6: Problems Faced in Seeking Information

CONCLUSION

Studying the information needs and seeking behaviour of faculty members in a library is the best way to understand their right information needs and this will also increase the effective use of library resources and services. The data analysis indicates that respondents use internet as their main source and channels for seeking information and e-books/e-journals are major information resources used. It has also been observed that library staff requires training regarding library resources and services. Moreover, the user education regarding library resources and services is also required. Seminars and workshops should be organized for the users also. This will facilitate the users to access the library resources in efficient and effective way.

REFERENCES

[1] Ajiboye and Tella (2007). University undergraduate students' information seeking behaviour: implications for quality in higher education in Africa. *The Turkish Online Journal of Educational Technology TOJET*, 6(1): 40–52.

[2] Choudhury, Tuhina (2013). Information Needs and Seeking Behaviour of Assam University Library Users. *IOSR Journal of Humanities and Social Science* (IOSR-JHSS), 11(6),10–23.

[3] Curtis, K. *et al.* (1997). Information-seeking behaviour of health sciences faculty: the impact of new information technologies. *Bulletin of the Medical Library Association*, 85(4), 402–410.

[4] Fiancor, D.K. and Adams, M. (2004). Information seeking behaviour and Information use by the students of the department of information studies, University of Ghana, Legon, *Ghana Library Journal*, Vol. 16, pp. 45–60.

[5] Hoshvosky, A.G. (1968). Information Source: Its ends, means and opportunities. *Information Transfer: Proceeding ASAS, 55(2),* 1968, 47–55. http://www.thefree dictionary.com

[6] Krikelas, J. (1983*)*. Information-seeking behaviour: Patterns and concepts. *Drexel Library Quarterly*, 19(2), 5–20.

[7] Natrajan, M. (2012). Information seeking behviour of students of Management Institutions in NCR of Delhi. *Trends in Information Management (TRIM)*, 8(2), 100–110.

[8] Ngozi *et al.* (2015). Information seeking behaviour of faculty members in a Nigerian University. *International Journal of Academic Research and Reflection*, 3(4), 95–99. www.idpublications.org

[9] Nnadozie (2008). The information needs of faculty members in a Nigerian private university: A self study. *Library Philosophy and Practice* (e-journal). https://digitalcommons.unl.edu/libphilprac/196

[10] Nwobasi, Rosemary N. *et al.* (2013). Information needs and seeking behaviour of students in two universities in IMO state, Nigeria. *Library Philosophy and Practice* (e-journal). http://digitalcommons.unl.edu/libphilprac.

[11] Patel, Urjita and Chaudhari, B.K. (2015). Information seeking behaviour of faculty members of the Agriculture Universities in Gujarat. *RET Academy for International Journals of Multidisciplinary Research (RAIJMR)*, 3(7), 20–26. http://www.raijmr.com/ijrhs/wp-content/uploads/2017/11/IJRHS_2015_vol03_issue_07_05.pdf

[12] Patitungkho, K. and Deshpande, N.J. (2005). Information seeking behaviour of faculty members of Rajabhat Universities in Bangkok. *Webology*, *2*(4), 25–33.

[13] Rafiq, Muhammad and Ameen, Kanwal (2009). Information Seeking Behavior and User Satisfaction of University Instructors: A Case Study. *Library Philosophy and Practice*. Retrieved from: https://digitalcommons.unl.edu/libphilprac/240

[14] Suriya, M. (2004). Information-seeking behaviour of faculty members from government arts colleges in Cuddalore district. In: Kaul, H.K. and Patil, S.K. (Eds), *Library and information networking (NACLIN 2004)*. New Delhi: 285–292.

[15] Thilagavathi, T. and Thirunavukkarasu, A. (2015). Information Needs and Seeking Behaviour of faculty members of Avinashilingam University, Coimbatore. *International Journal of Digital Library Services*, 5(3). 69–76. Retrieved from: http://www.ijodls.in/uploads/3/6/0/3/3603729/8535.pdf

[16] Wilson, T.D. (1981). On user studies and information needs. *Journal of Documentation*, 37 (1), 3–15.

[17] Wilson, T.D. (1997b). Information behavior: an inter-disciplinary perspective. *Information Processing & Management,* 33(40), 551–572.

[18] Wilson, T.D. (2000). Human Information Behavior. *Information Science,* 3(2) 49.

Epilogue

The purpose of this part of the book is to draw attention of the readers to key indicators given in each of the preceding chapters sequentially. The number coincides with the numbers of chapters in the book.

1. Blockchain keeps the record of the Contract which are executed by the network's distributed nodes, where the third-party requirement is eliminated as Blockchain is decentralized in nature. There are still multiple security and privacy issues with the Smart Contract which imposes a huge threat for its continued effectiveness and recognition.
2. The copper oxide nanoparticles can be synthesized by chemical co-precipitation method and the pure PANi and PANi-CuO nanocomposites can be successfully fabricated by chemical oxidative polymerization. The peaks obtained in XRD are in good agreement with the reported values. The characteristic peaks of FTIR and UV-Vis spectroscopy and the SEM images confirms the successful synthesis of materials. From the Four-probe conductivity measurements, it is clear that the conductivity increases when the pure PANi is converted into nanocomposites with CuO particles. The antimicrobial results of the nanocomposites recommend that the synthesized nanocomposites exhibit microbicidal effect.
3. The advancement in deep learning has encouraged to enhance task efficiency in computer vision. Detection efficiency has increased considerably, but still there is substantial variation in precision between normal and tiny objects. The disparity between small, medium, and large scales is too great in the COCO dataset parameters. Most models are effective at detecting normal objects, but when used to detect tiny objects, problems can be encountered.
4. Forensic analysis of ink plays a significant role in questioned document examination. The composition of ink is commonly kept confidential by the manufacturers and varies from brand to brand. It is possible to differentiate the handwriting written with different brands of ink with hyper spectral imaging (VSC-6000) to individualization of ball point pen ink of different manufacturers. Better characterization could be possible at 485-590 nm. Blank reference is suggested to be used in order to minimize spectral interferences.
5. The Online market is flooded with several, leading national and global brands besides numerous small and local brands. Online shopping, despite divergent brands, is not well differentiated by the consumers. This results in fragmented market and leads to a highly competitive market. A deep insight into brand loyalty and satisfaction can facilitate marketers to retain the

prevailing customers and stimulate new ones. Five factors namely Trust, Connect, Convenience, Fondness and Expectation can lead to the brand loyalty of online customers.

6. Precision agriculture automates farming, including gathering data and investigating for improved decision-making and to increase the yield and superiority of the cloud-based agricultural product. The CBA combined with data analytics and deep learning is so-called smart farming or smart agriculture. The objective of intelligent agriculture is to grow a decision-making provision system for farming management. The precision smart agriculture can be improved through the benefit of IoT's latest technologies and proposed Novel Deep Learning Algorithm.
7. Handwriting has been considered as an important tool for the identification of the suspect. Different attributes of handwriting have been useful in determining the gender of the individual. Handwriting is identified by the combination of all elements and characteristics related with all forms of letters obtained from the hand movement. Identification of handwriting is important evidence in the field of forensic science.
8. A person learns handwriting skills on a horizontal surface station (flat) with forearm movement because it is easiest for muscles of hand. Once muscles adopt the neuromuscular pattern of handwriting, writing becomes more smooth, speedy and perfect. But in some conditions writer can get inclined surface stations (angular/inclined) for writing. In such situations handwriting may be different from that on the horizontal surfaces. It is suggested that the writing surface area/station of the document under study should not be overlooked.
9. Emotional well-being refers to the emotional quality of an individual's everyday experience the frequency and intensity of experiences of joy, stress, sadness, anger, and affection - that makes one's life pleasant or unpleasant. Emotional well-being is regulated by a number of factors like income, age, work environment, stress, humour, resilience, social interaction and physical health amongst others. It is possible to improve one's emotional well-being through various ways like mindfulness, positive relationships and social support amongst others.
10. The event sentence classification model is based on event trigger and argument from news articles. Event trigger identification and event argument extraction is done by extensive analysis of unstructured news article data. To identify the event mention, ACE2005 Proposed event triggering words for various news articles can be used. Named entity recognizer (NER) and other local features can be used to identify event argument.
11. Every region has its own unique style and pattern of the primitive or folk art. These native arts of India are ethnic and simple and are distinctive enough to say a lot about its rich heritage. Folk arts are elementary paintings done on

walls and floor and uses a very basic graphic vocabulary like lines, circles, squares and triangles. These motifs or elements come from the nature around them, which is a result of man's close association with his immediate surroundings India has always been a subcontinent that portrayed traditional and cultural spirit through its multi- dimensional, conventional arts.

12. Many times a disputed document bears genuine signature but the actual person whose signature is on the document denies that he has ever signed such document. Forgery of this kind is prevalent in numbers of cases. The forger anyhow obtains a genuine signature either by request being known to the forger or by chance gets some blank paper which bears his genuine signature. After that the signed document is utilized by him to create forgery.

13 Lessing asserts that women who want freedom and equality in a patriarchal society need to come out of their traditional mind-sets. They need to be aware of their strengths and not undermine themselves as inferior to men. While doing their traditional roles they need to pursue their quest for freedom. Pursuit of freedom does not mean that they have to run away of their responsibilities towards the family and the society rather they have to realise that they have the capability of playing various roles simultaneously which elevates them to a status that is higher to men even in patriarchal society.

14. The fungal community was observed more in the petroleum polluted soils whereas, the growth of the fungal community at the controlled sites was seen within the normal range. Inferior growth of fungi like *Aspergillus* sp., *Penicillium* sp., *Fusarium* sp., *Trichoderma* sp., *Alternaria* sp., *Curvularia* sp., *Mucor* sp., *Rhizopus* sp., *Morssonina* sp., *Helminthosporium* sp, *Cladosporium* sp., and *Chaetomium* sp., was observed. But some of these fungal species can digest hydrocarbons such as the *Aspergillus* sp., *Penicillium* sp., *Fusarium* sp. These species increase the fertility of the soil by reducing the number of hydrocarbons in the soil.

15. Trevor depicts individuals with different perspectives on Anglo-Irish relations, although few characters share common background, but their perceptions, values and attitudes are usually very different from one another. His characters are ordinary citizens, not engaged in politics or conflicts. Still these Anglo-Irish relationships have an immense impact on people's lives, mostly due to the people among them who impose the notion of common guilt. The stories are set on the backdrop of the events of the 20th century, and they contain a reference to the real events and individuals. The self-reflective ending of the novel provides a bright contrast to pictures of misery and despair.

16. Modified MobileNet model requires much less memory in contrast to AlexNet and VGG16 models so MobileNet is selected as a light-weighted

CNN model to deploy into edge node for real time identification of diseases. The advantage of utilizing this approach is that there is no need to send samples of leaf images to the cloud or laboratory to analyse them. Despite being lightweight, MobileNet achieved the same accuracy which is 99.87% for both train and test images.

17. The astronomical clock contingent upon geographical territory is considered to produce data base. The input energy of road light varies according to the variations in natural light intensity obtained from the data base. After mid night the street activity is too less so framework enlightens lights with half intensity so this utilizes nearly half of the power. The framework utilizes real time clock, microcontroller and TRIAC based circuit for controlling input energy of a road light. This shows that the real time clock based framework is more reliable than sensor based frameworks.
18. Cannabis substance abuse is widely causing drug addiction problems across the world. In the drug abuse cases, hair plays an important role in identifying drug abuse and drug history consumption of the addict. However, scalp hair may not be available in every drug abuse case, but hair of other body parts play an essential role in forensic toxicological analysis.
19. The basic science will remain the need of the world always. The multi-disciplinary work approach needs to be enhanced amongst the students. Industry academia and research institutes have to work hand in hand for the development of the country. Domain knowledge along with professional and personality development is a need of the present day society. Faculty with domain knowledge and professional skills is the only way to bridge the gap. Government and institutes should take initiatives for developing skilled manpower.
20. The basic reproduction number, $\Re_0$ is the expected number of secondary infections a single infected individual can produce when introduced in population of susceptible. When $\Re_0$ is higher than one, disease outbreak will occur; when $\Re_0$ is lower than one disease often fails to spread and may remain endemic in population. In order to evaluate true condition more accurately, it is necessary to measure whether value of $\Re_0$ is higher or lower than one.
21. Prediction accuracy can be improved by focusing on both image analysis of facial expression and motion dynamics of eyes, lips, head and hand movement to compute time and space complexity. Deep learning algorithms generally process complex and large datasets containing a large number of features to get meaningful information. In most cases, these algorithms have high computational complexity issues while processing large datasets due to the presence of lesser weight features (irrelevant or redundant). As a result, the computing increases with increase in the number of features, number of records, feature types and inbuilt feature categories present in datasets.

22. Studying the information needs and seeking behaviour of faculty members in a library is the best way to understand their right information needs and this will also increase the effective use of library resources and services. Internet is the main source and channel for seeking information and e-books/e-journals are major information resources used. There is a need for training regarding library resources and services for the library staff. Moreover, the user education regarding library resources and services is also required. Seminars and workshops should be organized for the users also. This will facilitate the users to access the library resources in efficient and effective way.

Subject Index

Author Index

9 789387 997530

Printed at Repro India Ltd.